Introduction to the Theory of Boehmians

P.K. Banerji
Emeritus Professor
Principal Investigator, DST USERS
Department of Mathematics
J.N.V. University, Jodhpur, India

DST-USERS SCHEME BOOK SESIES
Department of Science & Technology
Ministry of Science & Technology, India

SCIENTIFIC
PUBLISHERS (INDIA)

Published by:

Scientific Publishers (India)
5 A, New Pali Road, P.O. Box 91
Jodhpur 342 001 (India)

E-mail: info@scientificpub.com
Website: www.scientificpub.com

Branch Office
Scientific Publishers (India)
4806/24, Ansari Road, Daryaganj
New Delhi - 110 002 (India)

ISBN: 978-81-7233-904-3

Printed in India

To My beloved Granddaughter

Ishita

Preface

Despite including our own research in the theory of Boehmians and Boehmian spaces and consulting and reviewing work of other researchers, it did not appear an easy task to write a moderate text in a very comprehensible manner. This monograph is an outcome of very many scattered research papers which are organized for a systematic reading. For comprehension and accessibility, collection of literature is made warily. In the preparation of the complete text, care is taken to simplify mathematics that is involved. The prerequisites for reading this book are real and complex analysis and to some extent, the functional analysis. We, at this juncture, may not prognosis the quantum of readers of this book, because we, in India (and for that matter in many parts of the world), do not run any course on Boehmian spaces. Moreover, this book is not a university text book. The aim of writing this book is to assist young researchers who plan their research in this direction.

I am grateful to many authors whose work have been freely consulted. I am thankful to Prem Lata Paliwal for the excellent job of typesetting the book and to Pukhraj Paliwal for Proof reading. I am thankful to the Department of Science and Technology, Government of India, New Delhi, India, for sanctioning (Sanction No. HR/UR/49/2010) the DST-USERS (Utilization of Scientific Expertise of Retired Scientist). Last but not the least, I am grateful to Pawan Kumar Sharma, Scientific Publishers (India) for granting to publish this book.

P.K. Banerji

Contents

Chapter 1

Schwartz Theory of Distributions

1.1 Prologue

The delta function $\delta(t)$, often called the impulse or Dirac delta function, occupies a significant importance in signal analysis. Most problems in physics are modeled as delta functions. Strictly speaking, delta function is not mathematically accepted function and it cannot be seriously treated within the framework of classical analysis. However, if distributions are introduced, then the concept of a delta function and operations on it can become meaningful.

1.2 Generalized Functions and Test Functions

A distribution is a generalization of a function (that, presumably, is the reason that distributions are called generalized functions). Within the framework of distributions, any function used (or come across) in applications, viz. unit-step functions and pulses, may be differentiated as many times as we desire, and any convergent series of functions may be differentiated term by term.

Suppose $f(x)$ represents temperature at a point x in a room or may be $f(x,t)$ is temperature at a point x and time t. We can measure the temperature with a thermometer, placing the bulb of it at the point x. Unlike the point, the bulb of the thermometer has a non zero size. Thus, what we measure is more an average temperature over a small region of space. Further, if we think of temperature as varying with time also, then we average over a small time interval preceding the time t when thermometer is actually read.

This in mathematical terms means that a thermometer measures

$$\int f(x)\varphi(x)dx, \tag{1}$$

where $\varphi(x)$ depends on the nature of the thermometer and where we keep it. $\varphi(x)$ will tend to be concentrated near the location of the thermometer bulb and will be nearly zero once we are sufficiently far away from the bulb. To say this is an average, is to require

$$\varphi(x) \geq 0 \qquad \text{everywhere} \tag{2}$$

and

$$\int \varphi(x)dx = 1 \, (\text{the integral is taken over all spaces}) \, . \tag{3}$$

Moreover, with two thermometers we can measure

$$\int f(x)\varphi_1(x)dx \tag{4}$$

and

$$\int f(x)\varphi_2(x)dx, \tag{5}$$

which when subtracted, deduces the value of

$$\int f(x)[\varphi_1(x) - \varphi_2(x)]dx, \tag{6}$$

where it can be noticed that $\varphi_1(x) - \varphi_2(x)$ is no longer non-negative. Proceeding further, we can even compute

$$\int f(x)[a_1\varphi_1(x) - a_2\varphi_2(x)]dx, \tag{7}$$

where a_1 and a_2 are constants, and $a_1\varphi_1(x) - a_2\varphi_2(x)$ may have any finite value for its integral.

This convinces us that it is often more meaningful physically to discuss quantities like $\int f(x)\varphi(x)dx$, than the value of f at a particular point x. Since to measure the value of f at x is essentially impossible, we stop requiring our functions to have a value at x. This means we are considering a larger class of objects, which are called generalized functions (we have answered the synonym here and in the beginning of this section). What we will require of a generalized function is, that something similar to $\int f(x)\varphi(x)dx$ exists for a suitable choice of averaging functions φ, call them test funtions.

Let us write $\langle f,\varphi\rangle$ for this something (we come back soon for more rigour discussions), which should be a real function (or may be a complex number in case we consider complex valued test functions and generalized functions). Let us recall

$$a_1\int f(x)\varphi_1(x)dx - a_2\int f(x)\varphi_2(x)dx$$
$$= \int f(x)[a_1\varphi_1(x) - a_2\varphi_2(x)]dx. \tag{8}$$

We want to do some sort of thing with generalized functions, so we should require

$$a_1\langle f,\varphi_1\rangle - a_2\langle f,\varphi_2\rangle = \langle f, a_1\varphi_1 - a_2\varphi_2\rangle. \tag{9}$$

This property is called linearity. If a_2 is replaced by $-a_2$, then (9) becomes

$$a_1\langle f,\varphi_1\rangle + a_2\langle f,\varphi_2\rangle = \langle f, a_1\varphi_1 + a_2\varphi_2\rangle\ . \tag{10}$$

We agree that if φ_1 and φ_2 are test functions, then $a_1\varphi_1 + a_2\varphi_2$ is also a test function. There are actually many possible choices for the collection of test functions, which lead to various theories of generalized functions. As has been mentioned earlier, we are back for more discussion on test functions. A test function $\varphi(t)$ is a real - valued function of the real variable that can be differentiated arbitrary number of times and is identical to zero outside a finite interval. For example, the test function

$$\varphi(t,a)=\begin{cases} e^{\frac{a^2}{a^2-t^2}} & ;|t|<a. \\ 0 & ;|t|\geq a. \end{cases} \quad (11)$$

Properties

1. If $f(t)$ can be differentiated, then

   $\psi(t)=f(t)\varphi(t)=$ test function

2. If $f(t)$ is zero outside a finite interval, then

   $$\psi(t)=\int_{-\infty}^{\infty} f(\tau)\varphi(t-\tau)d\tau, -\infty<t<\infty = \text{test function}$$

3. A sequence of test functions, $\{\varphi_n\}, 1\leq n<\infty$, converges to zero if all φ_n are identically zero outside some interval independent of n and each φ_n, as well as all of its derivatives, tend uniformly to zero. That is

   $$\varphi_n(t)=\varphi\left(t+\frac{1}{n}\right)-\varphi(t). \quad (12)$$

1.3 Properties of Distributions

1. Linearity – Homogeneity

$$\left.\begin{aligned} &\int_{-\infty}^{\infty} f(t)[a_1\varphi_1(t)+a_2\varphi_2(t)]dt \\ &\qquad = a_1\int_{-\infty}^{\infty} f(t)\varphi_1(t)dt + a_2\int_{-\infty}^{\infty} f(t)\varphi_2(t)dt \\ \text{i.e.}\quad &\langle f, a_1\varphi_1+a_2\varphi_2\rangle = a_1\langle f,\varphi_1\rangle + a_2\langle f,\varphi_2\rangle \end{aligned}\right\} \quad (13)$$

2. Summation

$$\int_{-\infty}^{\infty}[f_1(t)+f_2(t)]\varphi(t)dt$$

$$\left.\begin{aligned}&=\int_{-\infty}^{\infty}f_1(t)\varphi(t)dt+\int_{-\infty}^{\infty}f_2(t)\varphi(t)dt\\ \text{i.e.}\quad&\langle f_1+f_2,\varphi\rangle=\langle f_1,\varphi\rangle+\langle f_2,\varphi\rangle\end{aligned}\right\}\qquad(14)$$

3. Shifting

$$\int_{-\infty}^{\infty}f(t-t_0)\varphi(t)dt=\int_{-\infty}^{\infty}f(t)\varphi(t+t_0)dt\qquad(15)$$

4. Scaling

$$\left.\begin{aligned}&\int_{-\infty}^{\infty}f(at)\varphi(t)dt=\frac{1}{|a|}\int_{-\infty}^{\infty}f(t)\varphi\left(\frac{t}{a}\right)dt\\ \text{i.e.}\quad&\langle f(a),\varphi\rangle=\frac{1}{|a|}\left\langle f,\varphi\left(\frac{1}{a}\right)\right\rangle\end{aligned}\right\}\qquad(16)$$

5. Derivative

$$\left.\begin{aligned}&\int_{-\infty}^{\infty}\frac{d}{dt}f(t)\varphi(t)dt=-\int_{-\infty}^{\infty}f(t)\frac{d}{dt}\varphi(t)dt\\ \text{i.e.}\quad&\langle f',\varphi\rangle=\langle f,\varphi'\rangle\end{aligned}\right\}\qquad(17)$$

6. The *n* th Derivative

$$\left.\begin{aligned}&\int_{-\infty}^{\infty}\frac{d^n}{dt^n}f(t)\varphi(t)dt=(-1)^n\int_{-\infty}^{\infty}f(t)\frac{d^n}{dt^n}\varphi(t)dt\\ \text{i.e.}\quad&\left\langle f^{(n)},\varphi\right\rangle=(-1)^n\left\langle f,\varphi^{(n)}\right\rangle\end{aligned}\right\}\qquad(18)$$

7. Product with Ordinary Function

$$\left.\begin{aligned}&\int_{-\infty}^{\infty}[\alpha(t)f(t)]\varphi(t)dt=\int_{-\infty}^{\infty}f(t)[\alpha f(t)\varphi(t)]dt\\ \text{i.e.}\quad&\langle\alpha f,\varphi\rangle=\langle f,\alpha\varphi\rangle\end{aligned}\right\}\qquad(19)$$

where $\alpha(t)$ is the ordinary function, and the relation (19) is valid provided $\alpha(t)\varphi(t)$ belongs to the set of test functions.

8. Convolution

$$\int_{-\infty}^{\infty}\left[\int_{-\infty}^{\infty} f_1(\tau) f_2(t-\tau)d\tau\right]\varphi(t)dt$$

$$= \int_{-\infty}^{\infty} f_1(\tau)\left[\int_{-\infty}^{\infty} f_2(t-\tau)\varphi(t)dt\right]d\tau \tag{20}$$

9. A sequence of distributions $\{f_n(t)\}_{n=1}^{\infty}$ is said to converge to the distribution $f(t)$ if

$$\left.\begin{array}{l} \lim\limits_{n\to\infty} \int_{-\infty}^{\infty} f_n(t)\varphi(t)dt = \int_{-\infty}^{\infty} f(t)\varphi(t)dt, \\ \text{i.e.} \quad \lim\limits_{n\to\infty} \langle f_n, \varphi\rangle = \langle f, \varphi\rangle \end{array}\right\} \tag{21}$$

for all φ belonging to the set of test functions.

10. Every distribution is the limit, in the sense of distributions, of a sequence of infinitely differentiable functions.

11. If $f_n(t) \to f(t)$ and $g_n(t) \to g(t), g$ is a distribution, and the number $a_n \to a$, then

$$\left.\begin{array}{l} \frac{d}{dt} f_n(t) \to \frac{d}{dt} f(t), f_n(t) + g_n(t) \to f(t) + g(t), \\ a_n f_n(t) \to a\, f(t) \end{array}\right\} \tag{22}$$

12. Any distribution $f(t)$ may be differentiated as many times as desired. This mean, that the derivative of any distribution always exists and it is a distribution.

1.4 More on Distributions and Test Functions

The space $\mathcal{D}$ leads to the theory of distributions. We intend to describe other spaces of test functions. The underlying (basic/hidden/main) point set will be a n-variable space R^n, where $x=(x_1,x_2,...,x_n)$, or even a subset $\Omega\subset R^n$ that is open. This means that every point $x\in\Omega$ is surrounded by a ball $\{y:|x-y|<\varepsilon\}$ contained in Ω, where ε depends on x. As every open ball, $\Omega=R^n$ is open, intuitively, an open set is just a union of open balls.

The class of test functions $\mathcal{D}(\Omega)$ consists of all functions $\varphi(x)$ defined in Ω, becoming zero outside a bounded subset of Ω which is situated away from the boundary of Ω such that all partial derivatives of all orders of φ are continuous.

We now define the class of distributions on Ω, deonted by $\mathcal{D}'(\Omega)$, to be all continuous linear functionals on $\mathcal{D}(\Omega)$. By functional we mean a real (or complex) – valued function on $\mathcal{D}$, written as $\langle f,\varphi\rangle$ and satisfies the identity.

$$a_1\langle f,\varphi_1\rangle+a_2\langle f,\varphi_2\rangle=\langle f,a_1\varphi_1+a_2\varphi_2\rangle \ . \tag{23}$$

That is why we call it linear, see relation (13).

Note also, $a_1\varphi_1+a_2\varphi_2\in\mathcal{D}(\Omega)$ if $\varphi_1,\varphi_2\in\mathcal{D}(\Omega)$. By continuous, we mean that if φ_1 is close enough to φ, then $\langle f,\varphi_1\rangle$ is close to $\langle f,\varphi\rangle$, we write the exact definition later. Sobolev, in an article, defined functionals, which are finite order distributions. There are actually many possible choices for the collection of test functions, leading to many different theories of generalized functions.

Any function gives rise to a distribution by setting

$$\langle f, \varphi \rangle = \int_{\Omega} f(x)\varphi(x)dx,$$

at least if the integral can be defined. This is justified if *f* is continuous, but actually more general functions will work. Subject to the theory of integration in use, we may make *f* discontinuous and even unbounded, provided the improper integral converges absoluetly. For example, $\int_{|x|\le r} |x|^{-t}\, dx$ converges for $t < n$, (in *n–dim*) and diverges for $t \ge n$.

Hence $|x|^{-t}, t < n$, gives rise to the distribution in $\mathcal{D}'(R^n)$

$$\langle f, \varphi \rangle = \int_{R^n} \varphi(x)|x|^{-t}\, dx. \tag{24}$$

The answer to the question, with regard to the peculiar choice of test functions $\mathcal{D}$, is not very fluent. It may not be possible to use physical or some unexplained reasons to say $\mathcal{D}$ is the only natural class of test functions, but (on the other hand) it certainly produces an elegant and fruitful theory. We may feel that the smoother we require the test function φ to be, the rougher we can allow the generalized functions *f* to be. To define $\langle \delta, \varphi \rangle, \varphi$ must be at least continuous, and to define $\langle \delta', \varphi \rangle$, we require (essentially) φ to be differentiable.

The requirement that the test functions in $\mathcal{D}$ become zero outside a bounded set and near the boundary of Ω is less important. It allows distributions to grow rapidly in arbitrary manner, as we approach the boundary (or infinity). Later in the course, a smaller class of distributions is considerd, which is called tempered distributions, and cannot grow as rapidly at infinity by considering larger class of test functions that have weaker vanishing properties.

1.5 How Good are Distributions ?

Among many different properties of distributions, first is what we have in Section 1.2, their purpose is to solve problems of differentiation. The meaning of generalization is that the set of distributions is larger than the set of functions; every function is a particular distribution, but there are distributions, which are not functions. The need for such generalization arises owing to some problem in differentiation: there are many certainly differentiable functions, but Weierstrass was the first to give an example of a function which is nowhere differentiable. This example is, however, extreme, but we frequently come across functions which are differentiable at most points, but not differentiable at some exceptional points, for example the function $|x|$. Distributions solve the problems of differentiation in the sense that every distribution is differentiable and even infinitely differentiable, and differentials are also distributions.

It may be noted that if a continuous function has no derivatives, then considering it as a distribution, the function always has a derivative. This advantage made the use of distributions popular everywhere in calculus; they are always differentiable, even when derivatives in the usual sense do not exist. In the field of complex number every real number has two complex, opposite square roots, and we even obtain d'Alembert's theorem : *every polynomial with complex coefficients of degree m has m complex roots*. This is analogus to the relation between functions and distributions described above.

Peano wrote in 1912 on the difficulties of differentiation : he was sure that something must be found. He visioned the existance of notion of generalized functions, which are to functions what the real numbers are to rationals. This intuition arose long before 1944, when Schwartz discovered distributions. But the limitation of mathematical knowledge of that time did not make it possible for Peano to find the generalized functions.

1.6 Symbolic Calculus : Historical Note

In continuation to Section 1.5 (prior dealing with present), we try to answer the question. Reality models are constructed by using differential equations. The reality we are trying to model, sometimes, may propose an exclaimed suggestion that some solutions of the differential equation need not be differentiable. For instance, the vibrating string equation

$$\frac{\partial^2 u(x,t)}{\partial t^2} = k^2 \frac{\partial^2 u(x,t)}{\partial x^2} \tag{25}$$

has a solution $u(x,t) = f(z - kt)$ for any function of one variable *f*, which has the physical interpretation of a travelling wave with shape *f*(*x*) moving at velocity *k*.

There is no reason for the shape to be differentiable physically. In that case the equation (25) is not satisfied at some points. But then physically meaningful solution cannot be discarded owing to technical reasons. We may, therefore, become curious to think that if a function satisfies a differential equation except for some points where it is not differentiable, should it be admitted as a solution. We consider another instance.

Laplaces equation $\nabla u = 0$ is defined by

$$\nabla^2 u = \frac{\partial^2 u}{\partial x_1^2} + \frac{\partial^2 u}{\partial x_2^2} \quad in \ R^2 \tag{26}$$

and

$$\nabla^3 u = \frac{\partial^2 u}{\partial x_1^2} + \frac{\partial^2 u}{\partial x_2^2} + \frac{\partial^2 u}{\partial x_3^2} \quad \text{in } R^3 . \tag{27}$$

A solution to Laplace's equation has the physical interpretation of a potential in a region without charge. Calculations show that

$u(x_1,x_2)=\log(x_1^2+x_2^2)$ and $u(x_1,x_2,x_3) = (x_1^2+x_2^2+x_3^2)^{-1/2}$ are solutions at every point except at the origin, where they fail to be differentiable. But they must be rejected on physical grounds because potentials, away from charges, must be continuous, whereas these functions have poles.

The distribution theory allows us to distinguish between these two cases (mentioned above). In the first case, $u(x,t)=f(x-kt)$ as a distribution in R^2, satisfies the equation (25). Whereas, $u(x_1,x_2)=\log(x_1^2+x_2^2)$, as a distribution in R^2, does not satisfy the Laplace equation. In fact

$$\frac{\partial^2 u}{\partial x_1^2}+\frac{\partial^2 u}{\partial x_2^2}=c_1\,\delta \tag{28}$$

and in R^3, similarly,

$$\frac{\partial^2 v}{\partial x_1^2}+\frac{\partial^2 v}{\partial x_2^2}+\frac{\partial^2 v}{\partial x_3^2}=c_2\delta\,, \tag{29}$$

for $v(x_1,x_2,x_3)=(x_1^2+x_2^2+x_3^2)^{-1/2}$, for certain constants c_1 and c_2.

Now we discuss the symbolic calculus, but this note is brief. Heaviside, an electrical engineer, was the one who presumably preconceived the concept of distributions. He was keen to solve an ordinary differential equation with constant coefficient. Heaviside introduced, among already existing formulae to solve such equations, his symbolic calculus in 1893-94. Already, during that time, a composition product for function was in existance, which were zero outside of semi-axis [0,+∞), denoted by * (an asterisk), introduced by Mercer. For $t\geq 0$, the composition product is given by

$$(f*g)(t)=\int_0^t f(t-s)g(s)ds=\int_0^t f(s)g(t-s)ds\,, \tag{30}$$

which being both associative and commutative, defines an algebra. Composition plays a vital role in transitory phenomena (in electricity), that is, those which depend on time and start at time 0, so that they are represented by functions which are zero for $t \leq 0$. Composition, the word, was a misnomer, which is replaced by convolution, in English; in German, it is called Faltung. It has been defined and studied in hundred and thousand books and is present everywhere in analysis, in the study of functions which are zero on the negative real axis; it also occurs in the study of general locally compact topological groups, where the integral from 0 to t, in equation (30), is replaced by an integral over the entire group.

Heaviside introduced a symbolic calculus of convolution, where $f(t)$ is represented by this symbol, a function F of another variable p, and he wrote convolution multiplicatively

$$(f * g)(t) = F(p)G(p) . \tag{31}$$

Avoiding more detailed historical notes on Heavisde's work, we would like to conclude this section with an anecdote. Heaviside did plenty of other computations, with methods which were not acceptable in his time. A good mathematician of present time knows the justification of symbolic calculus, that in an algebra with unit and without zero divisors, one can, indeed, decompose a rational function into partial fractions. But was unimaginable for the analysts of Heaviside's time, and so his system was not accepted, despite the fact that the results were correct. Heaviside defended himself, but was defeated. He was demoralized by the failure of his theory to such an extant that he died mentally unstable. Laurent Schwartz wrote, *Some professions generally considered not to be dangerous, can unexpectedly become dangerous.*

1.7 Dirac Function and More

Dirac in 1926, for use in quantum physics, introduced the function, which carries his name and often called the Dirac measure, denoted by δ. But Heaviside defined it much earlier. Dirac gave an approximation of this function: for a very small number ε, he considered a function equal to $1/2\varepsilon$ in the interval $[-\varepsilon, \varepsilon]$, and said that as ε tends to 0, the function $1/2\varepsilon$ tends to the Dirac function δ. Indeed, in the theory of distributions, there is a topology implying a notion of convergence, for which it is true that Dirac's $1/2\varepsilon$ functions converge to the Dirac distribution δ. Dirac reasoned in the space R^N. He also gave another approximation of δ - function, namely the bell curve.

Moreover, even if several functions, depending on a parameter ε, converged to the Dirac δ- function as ε tends to 0, it was still important to see why, and then to define exactly what was meant by convergence. This was not done by Dirac. The Dirac distribution was actually justified well before the invention of distributions, as the Dirac measure. But like Heaviside, Dirac introduced the derivative of his function, whereas it is not possible to speak of the derivative of a measure. Dirac simply took the δ-function to be the limit of Gauss functions, and defined its derivative to be the limit of their derivatives; and a derivative can never be represented in this manner. Indeed, this is true that it (the derivative) has value 0 everywhere except at the origin, but then, to have meaning, it should also be 0 at the origin, since it is an odd function. It may be noted that the passage to the limit was later justified by the theory of distributions. Dirac and other physicist who worked with δ-function, went much farther to create a success of it. They made variable to change on their generalized functions.

For instance, they considered the function $\delta(u)$, where u was taken as the square of the hyperbolic Lorentz distance, $t^2 - x^2 - y^2 - z^2$. This has no meaning at all in function theory, but in relativistic quantum mechanics these functions were found to be useful. Eventually, all these formulae were later justified by the theory of distributions. Moreover, they were reproved with precise computations, on these singular functions, by Méthée, a Swiss mathematician, who published his work a few years after Schwartz published the distributions (1954).

The theory of harmonic and holomorphic functions are another necessary prerequisite for the theory of distributions. Since the time of Gauss, in the 19th century, it was known that a harmonic function can be defined due to the fact that it is continuous in space, and its average on every sphere is equal to its value at the centre. Therefore, the harmonicity of the function, which translates differentially into the fact that its Laplacian becomes zero, can be defined by a property which does not involve the existence of its derivatives. Holomorphy means that the function is differentiable with respect to the complex variable, and it can be characterized without reference to differentiation; moreover, a holomorphic function possesses successive derivatives of every order. In the theory of distributions, a continuous function u is called harmonic or holomorphic if ∇u *or* $\left(\frac{\partial}{\partial x} + i\frac{\partial}{\partial y}\right)u$ vanishes in the sense of distributions (Cauchy-Riemann relation).

Also this does not necessitate the existence of derivatives in the usual sense, and as mentioned above, it is a consequence of this property that a distribution satisfying these equations (the Cauchy – Riemann) is infinitly differentiable in the usual sense.

Plenty of other interesting historical and mathematical descriptions that I might have thought to include, are now decided to exclude. We have plenty of books on distributions and all of them contain extraordinary treasure and moreover, we decided to include some terminologies later in forthcoming chapters, as found appropriate. This book is proposed to deal with Boehmian and we, thus, decide to spend more time on it. For more reading, a constructive list of books are mentioned in Suggested Reading.

Chapter 2
Operational Calculus

2.1 Mikusiński's Operational Calculus

We may call this section a prologue to the theory of Boehmians and Boehmian spaces. Indeed, much care is required to write the entire text, that begins now. Several research papers have been written in between the time Thomas Kelman Boehme defended his doctoral thesis in 1960 and now. We are fortunate to have enrolled in this school of scholars working on Boehmians and its applications. Our aim is to write a comprehensive text on Boehmian by collecting scattered literature. We do not have, at this juricture of time, any claim as to how this treatise will appeal and about its conclave.

On July 27, 2012, twenty five years have passed since the demise of Jan Mikusiński, who belonged to the generation of outstanding Polish mathematicans and the creator of the operational calculus and the sequential theory of distributions. The participants of Professor Tadeusz Wazewski's seminar in 1943 were the first to come in contact with a new theory, which is now known, in the world of mathematics, as the Mikusiński operational calculus. Objects of the theory, operators, provide a common generalization of numbers and locally integrable functions on the positive half line. Professor Wazewski supervised Mikusiński's doctoral thesis, which he defended on 25 July 1945 at the Jagiellonian University in Cracow. On December 10, 1955, Mikusiński obtained the Doctor of Mathematics degree for a collection of papers under a title *A New Approach to the Operational Calculus*. Because of his eye disease he was forced to resign from teaching and research work in the middle of 1958. His application for a sick leave was rejected and therefore he had to quit his job at the Warsaw university, but he continued to be an employee of the Mathematical Institute of the Polish Academy of Sciences.

His arrival to Katowice (Silesia, Poland), a small town in southern Ploand, in 1960, is remembered as an event of great importance for the mathematicians working there, and shortly a group of young mathematicians from Higher Pedagogical School in Katowice got in touch with him. Professor Mikusiński delivered for them a series of lectures on operational calculus. After a year, the lectures turned to a regular seminar, first held at the School, then in Mikusiński's home. The informal scientific contacts of Professor Mikusiński with Silesian mathematicians became formal only in 1966, when the Mathematical Laboratory (now the Branch of the Mathematical Institute of Polish Academy of Sciences) was established. Professor Mikusiński was the head of the Mathematical Laboratory (IM PAS) till 1985.

The seminar led by Professor Jan Mikusiński, entitled *Generalized Functions and the Theory of Convergence*, was held in the seat of the Katowice Branch of the IM PAS. At the same time when operational calculus was advented, mathematicians took great interest in the theory of distributions, see [104, 69, 97, 98]. This theory was based on difficult concepts of functional analysis.

The first idea of symbolic operational calculus came from Leibnitz and to some extent by Euler, Laple, Cauchy among many others. Nevertheless, it is Oliver Heaviside who is regarded to be the father of the operational calculus, whose theory was first justified by Laplace transform. A total return to the original operator view point was made by Jan Mikusiński, who gave [71] entirely new foundations to the operational calculus, which he based on convolution quotients without referring it to the Laplace transform. L. Berg, T.K. Boehme, J.Burzyk, I.H. Dimovski, V.A. Ditkin were among many others who published work concerning operational calculus and its applications. The work Hypernumbers by Jan Mikusiński [71, 72] was written and published in 1944 in Poland under wartime conditions.

A pair of elements (α, f), consisting of a complex number α and vector f, belonging to an abstract space *W*, is called a hypernumber, which is written in the form

$$(\alpha, f) = \alpha + f\,. \tag{32}$$

The set of all possible pairs of complex numbers and vectors of the space *W* is called the hypernumber space and is denoted by [W]. A particular realization of the spaces [W]. is the set $\mathbb{C}$ of complex - valued continuous functions $f\ (t)$, defined for $t \in \langle 0, \infty \rangle$. The basis for the Mikusiński operational calculus is the Titchmarsh Theorem concerning the convolution of functions continuous on the positive halfline.

Let $\mathbb{C}$ denote the set of all real or complex valued continuous functions of real variable *t*, defined in the interval $0 < t < \infty$. If $f, g \in \mathbb{C}$, then by the convolution of f and g, denoted in the operational calculus by the symbol fg of the usual product, we mean the function

$$(fg)(t)=\int_0^t f(t-\tau)g(\tau)d\tau. \tag{33}$$

The Titchmarsh Theorem (in its weaker, so called, global version) states that if $fg=0$ in $[0,\infty)$, then $f=0$ on $(0,\infty)$ or $g=0$ on $(0,\infty)$, that is, the set $\mathbb{C}$ is the commutative ring with the usual addition and the convolution as multiplication without zero divisor. Therefore, this ring can be extended to a quotient field, i.e., to a set of fraction $\frac{f}{g}(g\neq 0)$ $(f,g\in\mathbb{C})$, where the divison denotes the converse operation to the convolution. The elements of this quotient field are called Mikusiński operators. This field is denoted by $\mathcal{M}$. Mikusiński's field contains all continuous functions f on $[0,\infty)$, because they may be represented by the fractions $f=\frac{f\cdot g}{g}$, where g is an arbitrary non-trivial continuous function on $[0,\infty)$. It contains, also, all elements α of the initial numerical field which can be identified with the functions $\frac{\{\alpha\}}{\{1\}}$. To distinguish the value $f(t)$ of the function at the point t from the function f itself, the symbol $\{f(t)\}$ is (sometimes) used, i.e., $\{\alpha\}$ and $\{1\}$ are constant functions equal to $\{\alpha\}$ and 1 on $[0,\infty)$.

According to the definition of convolution, we have

$$\{1\}\{f(t)\}=\left\{\int_0^t f(\tau)d\tau\right\}, \tag{34}$$

where the function $\{1\}$ is distinguished due to the formation of its convolution with arbitrary function $\{f(t)\}$ which causes the integration of the latter in the interval from 0 to t. Consequently, the function $\{1\}$ will be called the integral operator, denoted by l : $l=\{1\}$. It is easy to prove that

$$l^n = \left\{ \frac{t^{n-1}}{(n-1)!} \right\}. \tag{35}$$

A fundamental role in the operational calculus is played by the inverse of the integral operator $l=\{1\}$, denoted by $s=\frac{1}{l}$, which is called the differential operator. We have, by this definition, $ls=sl=1$.

2.2 Distributions : Sequential Approach

Later in the text we mention our research on the sequential approach of distributions. But it is perhaps a wise thinking to mention, at this stage, the concept of the sequential approach. We try to be reasonably brief without damaging the flavour.

The starting point for the definition of distributions in the sequential approach are sequences of continuous or smooth function (i.e., of class C^{∞}) in a certain fixed interval (a,b). Mikusiński's definition is analogous to that of real numbers in the Cantor's theory. A sequence $(\varphi_n)_{n\in N}$ of smooth (i.e., real-valued) functions in R^1 is said to be a fundamental sequence if for every finite open interval I in R^1 there exists a non-negative integer k and a sequence $(\Phi_n)_{n\in N}$ of smooth functions so that $\Phi_n^{(k)} = \varphi_n$ in I and $(\Phi_n)_{n\in N}$ uniformly converges in I. Two fundamental sequences $(\Phi_n)_{n\in N}$ and $(\Psi_n)_{n\in N}$ are equivalent, written as

$$(\Phi_n)_{n\in N} \sim (\Psi_n)_{n\in N}$$

if the sequence $\varphi_1, \psi_1, \varphi_2, \psi_2, \ldots$ is fundamental. The equivalence classes (formed of equivalence relation) are called distributions in R^1. An advantage of the sequential approach in the theory of distributions is the lucidity of extending to distributions many operations which are defined for smooth functions, i.e., regular operations. An example of regular operation is differentiation (of a given order $k \in N_0$) : $A(f) = f^{(k)}$, which can be performed for an arbitrary distribution *f*. We have seen in the preceding sections that every distribution is locally a distributional derivative of a finite order of a continuous function, whereas in the sequential approach it is a simple consequence of the definition given above and properties of the uniform convergence. It can be conveniently shown that the sequence $(f * \delta_n)$ is distributionally convergent to *f* for an arbitrary distribution $f \in R^1$, where δ_n is delta-sequence, i.e., a sequence of non-negative smooth functions such that $\int \delta_n = 1$ and $\delta_n(x) = 0$ for $|x| \geq \alpha_n$, $\alpha_n \to 0$ *as* $n \to \infty$.

In practice, operations are regular and non-regular. For example, the two-argument operations of product $A(\varphi, \psi) = \varphi.\psi$ and the convolution $A(\varphi, \psi) = \varphi * \psi$ are non-regular operations and they cannot be defined for arbitrary distributions. Mikusiński suggested a general method of defining irregular operations on distributions by using delta-sequences, see [2,73,74] and also refer to these for details.

Chapter 3

Generalized Schwartz Distributions : The Boehmians

3.1 Prologue : Definitions, Convergence

We prepare to learn Boehmian, we try to bring every possible information on Boehmian so as to furnish it to a reading book. May this book reach to all working on Boehmian, group of established researchers and group of young mathematicians who are proposing to begin a career in research in applicable functional analysis. Our own researches and those of others have been collected to structure this book and an endeavour is made to present an appropriate organised text. All those terminologies described in preceding two chapters will be frequently used.

In Heaviside's operational calculus, specifically with its application to partial differential equations, difficulties arose as a result of occurrence of certain operators suggested by Heaviside, justification of those were difficult and the range of the validity so developed remained unclear. The use of delta function and other impulse functions both in operational calculus and other branches of applied mathematics, presented similar lack of clarity with regard to the range of validity. Under such condition, in order to dissipate the rising confusion, one can either use operational calculus and impulse functions as a kind of heuristic means to formulate tentative solutions (which need verification) or else develop a mathematical theory to justify the process. Such a theory is the main objective of our present work, which is called convolution quotient, see Section 2.1.

3.2 The Delta Function

One often, mostly in physical sciences, deals with impulsive forces acting for a short duration. A unit impulse can be described by a function $p(t)$ which becomes zero outside a short interval and satisfies.

$$\int_{-\infty}^{\infty} p(t)dt = 1 \ .$$

It is convenient to idealize such forces as instantaneous and describe them by a function $\delta(t)$, which vanishes except for $t=0$. Such a function possesses the shifting property

$$\int_{-\infty}^{\infty} \delta(u) f(u) du = f(0) \tag{36}$$

for every continuous function *f*. Such, and similar other functions are used in mathematical physics, despite the existing fact that no such function can possess (in the sense of mathematics) the shifting property. The delta function being an idealization of function vanishing out side a short interval, it seems natural to approximate the delta function by such functions. If the function $p(t)$ satisfies

$$\int_{-\infty}^{\infty} p(t)dt = 1 \qquad \text{and} \qquad p(t) = 0$$

outside the interval (-1,1), then the function $p_n(t) = np(nt)$ satisfies the first of the above and becomes zero outside $(-1/n, 1/n)$ and may be regarded as an ever improving approximation to the delta function as $n \to \infty$. For all continuous function *f*, it can be shown that :

$$\lim_{n\to\infty} \int_{-\infty}^{t} p_n(u) f(u) du = f(0). \tag{37}$$

It may be noted that

$$\lim_{n\to\infty} \int_{-\infty}^{t} p_n(u) f(u) du = \begin{matrix} 0, t < 0 \\ 1, t \geq 0 \end{matrix} \tag{38}$$

shows that to some extent delta function may be regarded as the derivative of Heavisides unit function

$$H(t) = \begin{matrix} 0, t < 0 \\ 1, t \geq 0 \end{matrix}. \tag{39}$$

3.3 Locally Integrable Function

Let us consider a finite or infinite, or an open, or a closed or a half open (or half closed) interval *I* and a subset *S* of *I* consists of isolated points if the points of *S* have no limit point in *I*. In which case each closed finite subinterval of *I* will contain at most a finite number of points of *S*.

A function *f* is said to be locally integrable on the interval *I* if *f* (*x*) is defined on *I* with the possible exception of a set S_f of isolated points of *I*, which are called singularities of , if *f* (*x*) is continuous in *I* except at its singularities, and if the integral of *f* (*x*) over any closed subinterval of *I*, which contains no singularity of *f* (*x*) in its interior, exists either as a proper Riemann integral or else, as an absolutely convergent Riemann integral.

Let *f* be a locally integrable function in *I* and $[\alpha, \beta]$ be a closed subinterval of *I*. Suppose $c_1 < c_2 < ... < c_{n-1}$ are singularities of *f* in (α, β), and let $c_0 = \alpha$ and $c_n = \beta$ (may be singularities). Then for each k=0,1,2,...,n-1 each integral

$$\int_{C_k}^{C_{k+1}} f(x)dx$$

exists either as a proper Riemann integral or else as an absolutely convergent improper Riemann integral. We define

$$\int_{\alpha}^{\beta} f(x)dx = \sum_{k=0}^{n-1} \int_{C_k}^{C_{k+1}} f(x)dx, \tag{40}$$

and call $f(x)$ absolutely integrable. If I is a bounded closed interval, then a function which is locally integrable on I, is also absolutely integrable there. But in case of an infinite interval I, or if I is not a closed interval, then the above statement may not hold true. We may, for details, refer to Erdélyi [29].

3.4 Convolution Quotient

We consider R to be a field of complex numbers, t to be the non negative real variable, and $\mathbb{C}$ to be the class of real - or complex-valued continuous functions of a non-negative real variable.

Definition 1 : For $\alpha \in R$ and $a, b \in \mathbb{C}$, let $a+b$ is the function whose value at t is $a(t)+b(t)$; $\alpha\, a$ is the function whose value at t is

$$ab(t) = \int_0^1 a(u)b(t-u)du, \tag{41}$$

where ab is called the convolution of a and b. With regard to this definition, the algebraic properties are summarized as follows.

For $\alpha, \beta \in R$ and $a, b \in \mathbb{C}$, we have

(i) $a+b, \alpha a, ab \in \mathbb{C}$

(ii) $a+b = b+a, a+(b+c) = (a+b)+c$

(iii) $\alpha(a+b) = \alpha a + \alpha b, (\alpha+\beta)a = \alpha a + \beta a$

and $\alpha(ab) = (\alpha a)b = (a\alpha)b, (\alpha\beta)a = \alpha(\beta a)$,

(iv) $ab = ba, a(bc) = (ab)c, (a+b)c = ac + bc$.

Here, (i) defines $\mathbb{C}$ to be closed under the algebraic operations of addition, multiplication by a scalar, and the convolution; (ii) and (iii) define that $\mathbb{C}$ is a vector space over the field of complex numbers, the null element of it is $\{0\}$; (ii) and (iv) show that $\mathbb{C}$ is also a commutative ring, known as convolution ring, the zero element of which is the function $\{0\}$.

The convolution ring has no zero divisor, that is, if $a, b \in \mathbb{C}$, and $ab = \{0\}$, then either $a = \{0\}$ or $b = \{0\}$. This statement is a special case of the Titchmarsh Theorem. The absence of zero divisor in the convolution ring makes it possible to extend it (this ring) to a field, denoted by *F*, called the field of convolution quotients. Denote by *a,b,...* the continuous functions (i.e., elements of $\mathbb{C}$) and deonte by $\alpha, \beta, ...$ the numbers (i.e., elements of *R*).

Given $a, b, b \neq \{0\}$, the convolution equation $bz = a$ has at most one solution in $\mathbb{C}$; if z_1 and z_2 are both solutions, then

$$b(z_1 - z_2) = bz_1 - bz_2 = a - a = \{0\}, \tag{42}$$

which can be seen above, and since $\{b\} \neq 0$, $z_1 - z_2 = \{0\}$ or $z_1(t) = z_2(t), \forall t \geq 0$. But $bz = a$, has, infact, no solution in $\mathbb{C}$. We can now construct an extended algebraic system, a quotient field, where the equation $bx = a$, $b \neq \{0\}$, always has a solution. The construction is done in terms of ordered pairs (*a,b*) of continuous function, in which it is always assumed that $\{b\} \neq 0$, The ordered pair corresponds to the solution $bz = a$.

Definition 2: (i) Pairs (*a,b*) and (*c,d*) are equivalent if *ad* = *bc*, (ii) $\frac{a}{b}$ or –a / b is the class of all ordered pairs equivalent to (*a,b*). Each such class is called an equivalence class or a convolution quotient, and each ordered pair that belongs to *a* / *b* is called representative of *a/b*. It may be noted that (i) is reflexive, symmetric and transitive.

Definition 3 : (i) The set of all equivalence class is called the field of convolution quotients, which is denoted by *F*. (ii) The operations of addition, multiplication by a scalar, and convolution in *F* are defined, respectively, by

$$\frac{a}{b}+\frac{c}{d}=\frac{ad+bc}{bd},$$

$$\alpha\frac{a}{b}=\frac{\alpha a}{b}, \tag{43}$$

$$\frac{a}{b}.\frac{c}{d}=\frac{ac}{bd}.$$

3.5 Boehmians : Definition, Properties and Convergence

Among different versions of theory of distributions proposed, the first was due to Bochner [7] in which the Fourier transformation was generalized, and the second, the Mikusiński's operational calculus. Zemanian propounded [116, 117] on the basis of testing function spaces the concept of the theory of distributions, that contain appropriate kernel of integral transforms, and extended the classical theory of such integral transforms to the dual of testing function spaces, called generalized function spaces.

To generalize the notion of distributions and their properties, several attempts are made [2,18,17,35,106], but the theories differ from one another in generality and by their applications. Mikusiński's operational calculus (discussed in preceding sections) was another approach to the generalization of distributions, see Mikusiński [75]. The approach is algebraic, which involves the quotient field of the ring of all continuous functions which becomes zero for $x \le 0$ under addition and convolution. The problem arises is, that the Mikusiński operators are defined globally whereas their local properties are unknown. The other problem is, that the convergence structure, called Type I convergence (see Burzyk [13]), on the space of Mikusiński operators, is not topological.

Using an algebraic approach, similar to the contruction of Mikusiński operators, a new class of generalized functions, called Boehmians, was advented by Piotr Mikusiński. Thus, Boehmians and its spaces and types of the same are considered to be an excellent generalization of Schwartz theory of distributions. Indeed, the distribution space can be viewed as a proper subspace of the Boehmian space. Moreover, local properties of Boehmians are known; equality of two Boehmians on an open set can be defined, and the convergence structure given to Boehmians is topological, a complete metric totplogical vector space, see Mikusiński [76].

The concept of Boehmians is motivated by regular operators introduced by Boehme [9], see also Boehme [8]. Regular operators form a subalgebra of Mikusiński operators and hence, they include only such functions whose support is bounded from left. Boehmians possess an algebraic character of Mikusiński operators and simultaneously, do not have any restriction on the support.

General construction of Boehmians can be seen in [70,76], which when applied to various function spaces it describes many Boehmian spaces, containing all Schwartz distributions, Roumieu type ultradistributions, regular operators, and tempered distributions. The name Boehmian, as discussed above, is used for all objects through an abstract algebraic constructions, similar to the construction of the field of quotients. Let us consider G to be an additive commutative semigroup and $S \subseteq G$ is a sub-semigroup possessing a mapping * from $G \times S$ to G such that

(i) If $\delta, \eta \in S$, then

$(\delta * \eta) \in S$ and $\delta * \eta = \eta * \delta$

(ii) If $\alpha \in G$, δ, $\eta \in S$, then

$(\alpha * \delta) * \eta = \alpha * (\delta * \eta)$,

and

(iii) If $\alpha, \beta \in G$, $\delta \in S$, then

$(\alpha + \beta) * \delta = (\alpha * \delta) + (\beta * \delta)$.

The members of the class Δ, of sequences from S, are called the delta sequence, that satisfies

(i) If $\alpha, \beta \in G, (\delta_n) \in \Delta,$ and $\alpha * \delta_n = \beta * \delta_n, \forall n,$ then $\alpha = \beta$ in G,

(ii) If $(\delta_n), (\varphi_n) \in \Delta,$ then $(\delta_n * \delta_n) \in \Delta$.

Based on above facts, the class A of pair of sequences, defined by

$$A = \left\{ ((f_n), (\varphi_n) : f_n \subseteq G^N, (\varphi_n) \in \Delta \right\}, \tag{44}$$

where the elements of class A is called the quotient of sequences, doneoted by f_n / φ_n if

$$f_m * \varphi_n = f_n * \varphi_m, \forall m, n \in N. \tag{45}$$

Two quotients of sequences, f_n / φ_n and g_n / ψ_n, are said to be equivalent if

$$f_m * \psi_n = g_n * \varphi_m, \forall m, n \in N .$$

An equivalence relation on A and the equivalence classes are called Boehmians. The space of all Boehmians, usually deonted by β, is written as $x = f_n / \varphi_n$. Basic properties are

(i) Addition : $\left[\frac{f_n}{\delta_n}\right]+\left[\frac{g_n}{\varphi_n}\right]=\left[\frac{f_n*\varphi_n+g_n*\delta_n}{\delta_n*\varphi_n}\right]$

(ii) Scalar multiplication : $\alpha\left[\frac{f_n}{\delta_n}\right]=\left[\frac{\alpha f_n}{\delta_n}\right], \alpha\in C$

(iii) Multiplication : $\left[\frac{f_n}{\delta_n}\right].\left[\frac{g_n}{\varphi_n}\right]=\left[\frac{f_n*g_n}{\delta_n*\varphi_n}\right]$

(iv) Differentiation : $D^\alpha\left[\frac{f_n}{\varphi_n}\right]=\left[\frac{D^\alpha f_n}{\varphi_n}\right].$

In the Boehmian space β two types of convergences, viz. $\delta-$ convergence and $\Delta-$ convergence are studied, which are defined as under:

Definition 4 : A sequence of Boehmians $\{x_n\}$ in β is said to be $\delta-$ convergent to a Boehmian x in β, which is denoted by $x_n \xrightarrow{\delta} x$ if there exists a delta sequence $(\delta_k)\in\Delta$ such that

$$(x_n*\delta_k).(x*\delta_k)\in G, \forall k,n\in N \tag{46}$$

and

$$(x_n*\delta_k)\to(x*\delta_k) \text{ as } n\to\infty \text{ in } G, \forall k\in N. \tag{47}$$

Definition 5 : A sequence of Boehmians $\{x_n\}$ in β is said to be $\Delta-$ convergent to a Boehmian x in β, that is denoted by $x_n \xrightarrow{\Delta} x$ if there exists a delta sequence $(\delta_n)\in\Delta$ such that

$$(x_n-x)*\delta_n\in G,\ \forall n\in N \tag{48}$$

and

$$(x_n-x)*\delta_n\to 0 \quad \text{as } n\to\infty \text{ in } G.$$

In particular, if $G=(C^\infty(R))$, the space of infinitely differentiable functions on R, which is equipped with the topology of uniform convergence of functions, and all derivatives on compact sets and delta sequence defined, then a sequence of functions $\varphi_n \in \mathcal{D}$ such that

(i) $\int \varphi_n(x)dx = 1, \forall n \in N$

(ii) $\int |\varphi_n(x)| \leq C$, *for all* $n \in N, C$ being some constant

(iii) supp $\varphi_n \to 0$ as $n \to \infty$,

we get Boehmian space $\beta(C^\infty(R), \Delta)$, members of which are called C^∞ – Boehmians.

Whereas if G is the set of all locally integrable functions on R and identify two such functions (if they are equal almost everywhere) with respect to the Lebesgue measure, then the topology of this space is considered to be the semi-norm topology, generated by

$$p_n(f) = \int_{-n}^{n} |f| d\lambda, \quad n = 1,2,3... \tag{49}$$

where λ is the Lebesgue measure on R and D (R), and Δ are the classes of sequences. Then we obtain a corresponding Boehmian space $\beta(G, \Delta)$, called the space of locally integrable Boehmians, see details in Loonker [60], Banerji and Loonker [3], and Nemzer [89]. If we consider $G = C(R^N)$ and the delta sequences are defined as a sequence of functions $\varphi_n \in \mathcal{D}$ (under usual conditions), then the Boehmian space, so obtained, contains properly the space $\mathcal{D}'$ of Schwartz distributions.

Chapter 4

Wavelet Transforms, Fractional Calculus and Integral Transforms

4.1 Abridged Historical Indroduction

Although wavelet analysis is a relatively young mathematical subject, it has drawn attention, not only among mathematicians themselves, but from various other disciplines as well. In this context Chui [16] writes, *it is fair to attribute the main driving force of the rapid development of this field to the users rather than to the inventors of mathematics*. To the mathematicians, Fourier analysis has remained a very important research area, whose theory is nice, techniques are powerful, and its impact on science and technology most profound. However, even as early as the decade of the 1940s, those who used the Fourier approach to analyze natural behaviours were already frustrated with the limitiation of the Fourier transform and Fourier series in the investigation of physical phenomena with non-periodic behavior and local variations. The need for simultaneous time frequency analysis led to the introduction of Gabor's short time Fourier transform in 1946 and the Wigner-Ville transform in 1947. The common ingredient of these two transforms is the sinusoidal kernel in the core of their definitions such that both high and low-frequency behaviours are investigated in similar manner and any signal under investigation is matched by the same rigid sinusoidal wave form. In place of sinusoidal kernel as modulation (for phase shift), a French geophysicist, Jean Morlet et al. [80,81] introduced the operation of dilation, while retaining the translation operation, and developed an algorithm for the recovery of signals under investigation from this wavelet transform (in other words, he first introduced the idea of wavelet transform as a new mathematical tool for seismic signal analysis).

It was the mathematical physics group in Marseille, led by Alex Grossman, in association with Ingrid Baubechies and T. Paul; and others, extended Morlet's discrete version of wavelet transform to the continuous version, by relating it to the theory of coherent states in quantum physics. This is how the notion of the integral (or continuous) wavelet transform was introduced. Grossman, having made a sharp recognition of the importance of the Morlet wavelet transform, developed an exact inversion formula for the wavelet transform. Grossman and Morlet [37] made extensive study of the continuous wavelet transforms and their applications. It has then become clear that, analogous to the Fourier expansions, the wavelet theory has provided a new method for decomposing a function or signal. In order to eliminate the weakness of the Fourier transform (that it does not contain any local information), Gabor [33], a Hungarian British physicist and engineer, first introduced the Windowed-Fourier transform (or the short-time Fourier transform), or more appropriately the Gabor transform by using a Gaussian distribution function as the window function.

The idea is to use a window function in order to localize the Fourier transform, and then shift the window to another position and so on. The remarkable feature of the Gabor transform is that the local properties of the Fourier transform with the time resolution is equal to the size of the window. However, the Gabor wavelets suffer from some serious algorithmic difficulties which have successfully been resolved by Henrique Malvar [65, 66], by introducing new wavelets, now known as Malvar wavelets and fall within the general framework of window Fourier analysis. From an algorithmic point of view, the Malvar wavelets are much more effective and superior to other wavelets including Gabor wavelets and Grossman-Morlet wavelets.

Yves Meyer [68], a French mathematician, recognized the depth of the connection between the Calderòn formula in harmonic analysis [69] and the new algorithm discovered by Morlet and Grossman. Using the concept of the Calderon-Zygmund operators and the Littlewood-Paley theory, Meyer was able to give a mathematical foundation for the wavelet theory. Then came a turning point in the world of wavelets due to Daubechies et al. [22], who suggested a new construction of painless non-orthogonal wavelet expansion. Further, the work of Lemaire and Meyer [48] on the construction of a smooth orthonormal wavelet basis on R and R^n marked the beginning of their famous contributions to the wavelet theory.

Simultaneously, Stéphans Mallat recognized that some quadratic mirror filters play an important role for the construction of orthogonal wavelet bases generalizing the Haar system. Meyer [68] and Mallat [64] realized that the orthogonal wavelet bases could be constructed systematically from a general formalism [61,63], known as multiresolution analysis. Again, Mallat [62] constructed the wavelet decomposition and reconstruction algorithms using the multiresolution analysis. Some times later, Battle [6] and Lemaire [49] independently, proposed the construction of spline orthogonal wavelets with exponential delay.

Inspired by the work of Meyer, Ingrid Daubechies [20] propounded to wavelet theory by constructing families of compactly supported orthonormal wavelets with some degree of smoothness. This paper brought revolution in the study of wavelets and their diverse applications, by explaning the connection between the continuous wavelets on R, and the discrete wavelets on Z or Z_n, the latter is useful for digital signal analysis.

The idea of frames was introduced by Duffin and Schaeffer [28] and subsequently studied by Daubechies [19,21]. We conclude this brief note and suggest for further reading Debnath [23,24,25], Dikshit [27], Resnikoff and Wells [94], Walker [108], Rao [93], Ojha [90] among others.

4.2. Wavelets : Terminologies

Process of communication requires representation and analysis of a signal or an image, viz. a musical note, an electrical signal, and satellite images. One of the classical tools in representing and analyzing analog and digital signals is Fourier analysis, in which a signal is represented as a superposition of waveforms. The components of various waves in the signal are explicitly specified by using the orthogonality of these waveforms. The sinusoidal waves work as basic building blocks. The space $L^2(0,T)$ of Lebesgue measurable functions are square integrable over (0,*T*). We know that the set $\{e_n\}_{-\infty}^{\infty}$, for $e_n(t) = e^{inw_0t}$ forms an orthonormal basis of $L^2(0,T)$, where $w_0 = 2\pi / T$, is called the fundamental frequency. Thus, we express $f \in L^2(0,T)$ as

$$f(t) = \sum_{j=-\infty}^{\infty} c_j e^{inw_0 t}, \tag{50}$$

where

$$c_j = \langle f, e_j \rangle = \int_0^T f(t) e_{-j}(t) dt,$$

the inner product of *f* with *g* is denoted by $\langle f, g \rangle$. Representation (50) is known as Fourier series expansion of *f*. Since each e_j is a combination of sinusoidal waves sin *jx* and cos *jx*, we may state that the Fourier series expansion of *f* is equivalent to resolving the function in terms of its various frequency components nw_0. It may be noted that this spectrum of *f* (description of amplitudes at various frequency levels) exists only at discrete values of frequencies, i.e., at $\pm w_0, \pm 2w_0, \ldots,$ etc.

It is, hence, a discrete spectrum, also referred to as line spectrum. Due to this reason that equation (50) is called a discrete Fourier transform. Let, we presume, a signal is defined on an infinite time domain, let on whole of *R*. If the signal has a periodic behaviour, then the Fourier series expansion can be defined on the whole of the real line, by extending its definition periodically. For the non-periodic case, however, the Fourier expansion does not work, because the function $\{e_n\}$ does not belong to $L^2(R)$, space of Lebesgue measurable square integrable function over *R*.

Fourier transform is a powerful tool in representing functions in time domain and in their frequency domain. But the formula fails in transforming a signal from time domain to the frequency domain. Another drawback of the Fourier transform is that the formula does not provide any information about frequencies which evolve with time. In view of the above drawbacks, Gabor (1946), introduced a time localization method by using a window function in the definition of the transform.

Definition 1: A function g is called a window function if $g \in L^2(R)$ and $xg \in L^2(R)$.

Definition 2 : If g is a window function in $L^2(R)$, then its centre and width are defined by

$$t^* = \frac{1}{\|g\|_2^2} \int_{-\infty}^{\infty} x|g(x)|^2 dx, \tag{51}$$

and $2\Delta_g$, respectively, and the radius is given by

$$\Delta_g = \frac{1}{\|g\|_2} \left\{ \int_{-\infty}^{\infty} (t-t^*)^2 |g(t)|^2 dt \right\}^{1/2}. \tag{52}$$

Here $\|\cdot\|_2$ denotes the 2-norm of $L^2(R)$, given by

$$\|g\|_2 = \left\{ \int_{-\infty}^{\infty} |g(x)|^2 dx \right\}^{1/2}. \tag{53}$$

Geometrically speaking, by the centre of a window function, we mean the value in the time domain around which g has maximum spectral energy; the width of the window function specifies that the energy contribution outside the interval $[t^* - \Delta g, t^* + \Delta g]$ will be almost negligible.

Definition 3 : Let $\alpha > 0$ be fixed. Then the Gabor transform of $f \in L^2(R)$ is defined by

$$\left(G_\beta^\alpha f\right)(w) = \int_{-\infty}^{\infty} e^{-iwt} f(t) g_\alpha(t-b) dt, \tag{54}$$

where $b \in R$ and and g_α is defined by

$$g_\alpha(t) = \frac{1}{2\sqrt{\pi\alpha}} e^{-t^2/4\alpha} \tag{55}$$

Gabor transform localizes the Fourier transform around the point $t=b$. Since

$$\int_{-\infty}^{\infty} g_\alpha(x)dx = 1, \tag{56}$$

we have

$$\int_{-\infty}^{\infty} \left(G_\varphi^\alpha f\right)(w)db = \hat{f}(w) \ . \tag{57}$$

Hence, the Gabor transform decomposes the Fourier transform $\hat{f}$ of f to give its local spectral information.

The identity,

$$\left(G_b^\alpha f\right)(w) = \frac{e^{-ibw}}{2\sqrt{\pi\alpha}} \left(G_w^{1/4\alpha} \hat{f}\right)(-b),$$

i.e.

$$= \sqrt{\frac{\pi}{\alpha}} e^{-ibw} \frac{1}{2\pi} \int_{-\infty}^{\infty} \varepsilon^{ibn} \hat{f}(\eta) g_{1/4\alpha}(\eta - w)d\eta \tag{58}$$

shows that the window Fourier transform of f with window function g_α, at $t=b$, agrees with the window inverse Fourier transform of $\hat{f}$ with window function $g_{1/4\alpha}$ at $\eta = w$. In Gabor transform or the short time Fourier transform, the width of the window remains unchanged. Window function be such that the spectrum of the signal can be analysed locally more efficiently. This, in other words, would mean that the width of the window should be relatively small for high frequency levels and be wide enough to cover low levels. This concept generates the explanation and the definition of continuous wavelet transform.

Let there is a window function $\psi \in L^2(R)$ such that its Fourier transform $\hat{\psi}$ is also a window function. Let t^*, $2\Delta_\psi$ and w^*, $2\Delta_{\hat{\psi}}$, respectively, are the centers and widths of ψ and $\hat{\psi}$.

We define, for $a, b \in R, a \neq 0,$

$$\psi_{b,a}(t) = |a|^{-1/2} \psi\left(\frac{t-b}{a}\right). \tag{59}$$

The continuous wavelet transform of a signal $f \in L^2(R)$ may be defined as

$$\begin{aligned}(W_\psi f)(b,a) &= \langle f, \psi_{b,a} \rangle \\ &= \int_{-\infty}^{\infty} f(t) \quad \overline{\psi_{b,a}(t)dt.}\end{aligned} \tag{60}$$

Introduction of parameter a allows $\psi_{b,a}$ to compress or expand subject to the choice of a. Normalizing factor $|a|^{-1/2}$ automatically takes care of the reative frequency bond. For instance, if $a > 1$, , then $\psi_{b,a}$ is stretched by a factor a in the horizontal direction, whereas if $0 < a < 1$, then it is compressed in the same horizontal direction. The factor $|a|^{-1/2}$ reduces its size in the vertical direction and for small a, with the compression of $\psi_{b,a}$ in the horizontal direction, ψ is enlarged in the vertical direction. In fact, the total energy of $\psi_{b,a}$ remains independent of b and a, i.e.,

$$\|\psi_{b,a}\|^2 = \|\psi\|^2 \tag{61}$$

Further, we have

$$(W_\psi f)(b,a) = \langle f, \psi_{b,a} \rangle = \frac{1}{2\pi} \langle \hat{f}, \hat{\psi}_{b,a} \rangle, \tag{62}$$

and also, we note that

$$\frac{1}{2\pi}\hat{\psi}_{b,a}(w) = \frac{|a|^{-1/2}}{2\pi}\int_{\mathbb{R}} e^{-iwt}\psi\left(\frac{t-b}{a}\right)dt$$

$$= a|a|^{-1/2}\, e^{-ibw}\, \hat{\psi}(aw). \tag{63}$$

By setting $g(w) = \hat{\psi}(w+w^*)$, we can shift the centre of the window function to the origin. Employing (62) in the relation (63), we get

$$(W_\psi f)(b,a) = \frac{a|a|^{-1/2}}{2\pi}\int_{-\infty}^{\infty} f(w)e^{ibw} g(aw-w^*)dw. \tag{64}$$

Width of $g(aw-w^*)$ is given by $\frac{2}{a}\Delta_{\hat{\psi}}$. Hence, the frequency window, in this case, is given by

$$\left[\frac{w^*}{a} - \frac{1}{a}\Delta_{\hat{\psi}}, \frac{w^*}{a} + \frac{1}{a}\Delta_{\hat{\psi}}\right]. \tag{65}$$

As a consequence, therefore, the rectangular time frequency window is given by

$$\left[b + at^* - a\Delta_\psi, b + at^* + a\Delta_\psi\right]$$

$$\times\left[\frac{w^*}{a} - \frac{1}{a}\Delta_{\hat{\psi}}, \frac{w^*}{a} + \frac{1}{a}\Delta_{\hat{\psi}}\right]. \tag{66}$$

For explicit analysis, we need a reconstruction formula for functions (i.e., signals) from frequency domain to time domain. In what follows now is a brief note on the evaluation of the inverse wavelet transform.

We write

$$\tilde{f}(a,b) = (W_\psi f)(b,a) = \langle f, \psi_{b,a}\rangle$$

$$= \frac{1}{2\pi}\langle \hat{f}, \hat{\psi}_{b,a}\rangle,$$

which, when (63) is used, reduces to

$$\tilde{f}(a,b) = \frac{a|a|^{-1/2}}{2\pi} \int_R e^{ibw} \overline{\hat{\psi}(aw)} \hat{f}(w) dw,$$

i.e.
$$= a|a|^{-1/2} \left(\hat{f}(w) \overline{\hat{\psi}(aw)} \right)^{\vee} (b), \tag{67}$$

in which $\breve{f}(b)$ denote inverse Fourier transform of *f* at *b*.

Now apply Fourier transform on both sides of (67), multiply the resulting relation by $\hat{\psi}(aw)$, and then integrating with respect to *a*, we obtain $\hat{f}(w)$. Let, for the moment, *a* to be positive. Then

$$\int_0^\infty q(a) \hat{\psi}(aw) \int_{-\infty}^\infty \tilde{f}(a,b) e^{-iwb} dbda$$

$$= \left[\frac{1}{2\pi} \int_0^\infty q(a) a^{1/2} |\hat{\psi}(aw)|^2 da \right] \hat{f}(w) \tag{68}$$

Set

$$Y(w) = \int_0^\infty q(a) a^{1/2} |\hat{\psi}(aw)|^2 da, \tag{69}$$

and assume that there exists constants *A* and *B* such that $0 < A \le Y(w) \le B < \infty$, almost everywhere. Then

$$\hat{f}(w) = 2\pi \int_0^\infty \frac{q(a)}{\sqrt{a}} \int_{-\infty}^\infty \hat{\psi}^{b,a}(w) \tilde{f}(a,b) dbda, \tag{70}$$

where

$$\hat{\psi}^{b,a}(w) = \frac{1}{Y(w)} \hat{\psi}_{b,a}(w)$$

i.e.
$$= \frac{1}{Y(w)} a^{1/2} \hat{\psi}(aw) e^{-iwb} . \tag{71}$$

Hence f may be obtained as superposition of new functions $\psi^{b,a}$ as elements in $L^2(R)$. We can notice (details are avoided) that $\hat{\psi}^{b,a} \in L^2(R)$ with

$$\left\|\hat{\psi}^{b,a}\right\| \leq A^{-2}\|\hat{\psi}\|^2,$$

where,

$$\hat{\psi}^{b,a}(u) = a^{1/2}\int_{-\infty}^{\infty}(Y(w))^{-1}\hat{\psi}(aw)e^{iw(u-b)}dw$$

$$= \psi^a(u-b), \quad \text{say} \tag{72}$$

where

$$\psi^a(u) = a^{1/2}\int_{-\infty}^{\infty}(Y(w))^{-1}\hat{\psi}(aw)e^{iwu}dw$$

$$= a^{1/2}\int_{-\infty}^{\infty}\left(Y\left(\frac{w}{a}\right)\right)^{-1}\hat{\psi}(w)e^{iwu/a}dw \ . \tag{73}$$

By using (72), a sufficient condition, for the new family to have a single generating function, is

$$Y\left(\frac{w}{a}\right) = Y(w), \tag{74}$$

for almost all $w \in R$ and $a > 0$. We choose the weight function q so that Y satisfies (74). We need $q(a)a^{1/2} = 1/a$, to obtain

$$Y(w) = \int_0^{\infty}\frac{|\hat{\psi}(aw)|^2}{a}da \ = \ \int_0^{\infty}\frac{|\hat{\psi}(\xi)|^2}{\xi}d\xi = C_{+,} \text{, say} \quad (w > 0) \tag{75}$$

Also, for $w < 0$,

$$Y(w) = \int_0^{\infty}\frac{|\hat{\psi}(aw)|^2}{a}da \ = \ \int_0^{\infty}\frac{|\hat{\psi}(-\xi)|^2}{\xi}d\xi = C_{-}\text{, say}. \tag{76}$$

Thus Y is bounded if and only if $0 < C_{\pm} < \infty$. This means that we require

$$\int_{0}^{\infty} \frac{|\hat{\psi}(\pm\xi)|^2}{\xi} d\xi < \infty. \tag{77}$$

This is known as admissibility condition. We may write following :

Theorem 1 : Let $\psi \in L^2(R)$ be such that

$$C = \int_{-\infty}^{\infty} \frac{|\hat{\psi}(\xi)|^2}{\xi} d\xi \quad < \infty \; ; \qquad \text{[cf. Eqn.(77)].}$$

Then $f \in L^2(R)$ can be recovered from its wavelet transform $\tilde{f}$ by

$$f(t) = \frac{1}{C} \int_{-\infty}^{\infty} \int_{-\infty}^{\infty} a^{-2} \psi^{b,a}(t) \tilde{f}(a,b) dbda. \tag{78}$$

This is called the reconstruction formula.

Definition 4 : Let $\psi \in L^2(R)$ be such that

$$C_{\Psi} = \int_{-\infty}^{\infty} \frac{|\hat{\psi}(w)|^2}{w} dw < \infty. \tag{79}$$

Then ψ is called a basic wavelet (or mother wavelet).

The function $\psi^1 \in L^2(R)$, defined by

$$\psi^1(t) = \left(\frac{2}{C_{\psi}} \right) \psi(t) \tag{80}$$

is called the dual mother wavelet of ψ, viz. the family $\{\psi^{b,a}\}$ is called dual wavelet family to $\{\psi_{b,a}\}$.

Notice that the admissibility condition (77) implies that $\hat{\psi}(w) \to 0$ as $w \to 0$. If $\hat{\psi}$ is continuous, then $\hat{\psi}(0) = 0$, that suggests

$$\int_{-\infty}^{\infty} \psi(t) dt = 0. \tag{81}$$

This equation leads to the nomenclature Wavelet to such function.

4.3 Fractional Calculus

Phillip J. Davis mentioned that derivatives of arbitrary order are defined in terms of the gamma function in his Chauvenet prize winning paper in 1959. Fascination of many authors with the concept of interpolating between integral orders of derivatives led each of them to a serious study of what is sometimes called fractional calculus. During the first international conference on fractional calculus and its applications, held at the University of New Haven, West Haven, USA, in 1974, the name fractional calculus was considered a misnomer and decided it to be labelled generalized integration and differentiation.

Our dealing with this branch, in the present text is limited to distribution spaces and Boehmians. However, we will not escape necessary definitions. Fractional calculus, during the past four decades, has invited non-homogeneous scholars, viz. mathematicians, engineers, applied mathematicians, chemists and educators. The original question that led to the name fractional calculus was that, *can the meaning of a derivative of integer order* $d^n y / dx^n$ *be extended to have meaning when* n *is a fraction* ? Later this question was reframed, *Can n be any number : fractional, irrational or complex* ? Since the latter question was answered affirmatively, the name fractional calculus has become misnomer.

Euler in [30] wrote : *"When n is a positive integer, and if p be a function of x, the ratio* $d^n p$ to dx^n *can always be expressed algebraically, so that if* $n = 2$ *and* $p = x^3$, *then* $d^2 x^3$ *to* dx^2 is 6x to 1". It was asked, what kind of ratio can then be made if n is a fraction. Lagrange [45], developed the law of exponents (indices) for differential operators of integer order and mentioned

$$\frac{d^m}{dx^m}.\frac{d^n}{dx^n}y \quad = \frac{d^{m+n}}{dx^{m+n}}y. \tag{82}$$

The dot (.) is omitted (in modern notation), because it is not a multiplication. Laplace [46] defined a fractional derivative by means of an integral and the first mention of a derivative of arbitrary order appears in a text.

Lacroix developed a mere mathematical exercise generalizing from a case of integer order. With $y = x^m, m$ being a positive integer, he gave the relation.

$$\frac{d^n y}{dx^n} = \frac{m!}{(m-n)!}x^{m-n}, m \geq n \tag{83}$$

By invoking the generalized factorial (the gamma function), he obtained

$$\frac{d^n y}{dx^n} = \frac{\Gamma(m+1)}{\Gamma(m-n+1)}x^{m-n}, \tag{84}$$

where $\Gamma(n) = n!$ and $\Gamma(n+1) = n\Gamma(n)$. For $y = x$ and $n = 1/2$, he obtained

$$\frac{d^{1/2}y}{dx^{1/2}} = \frac{2\sqrt{x}}{\sqrt{\pi}}. \tag{85}$$

Fourier [3] was next to mention derivatives of arbitrary order, the definition of which was obtained from his integral representation of $f(x)$:

$$f(x) = \frac{1}{2\pi}\int_{-\infty}^{\infty} f(\alpha)d\alpha \int_{-\infty}^{\infty}\cos\, p(x-\alpha)dp. \tag{86}$$

The first use of fractional operations was made by Abel [1], while applying the fractional calculus in the solution of an integral equation that arises in the formation of the tautochrone problem. The integral equation is

$$k = \int_0^x (x-t)^{-1/2} f(t)dt, \tag{87}$$

where except for the multiplicative factor $1/\Gamma\left(\frac{1}{2}\right)$, equation is a particular case of a definite integral that defines fractional integration of order ½. Abel wrote the right hand side of (87) as

$$\sqrt{\pi}\left[d^{-1/2} / dx^{-1/2}\right] f(x).$$

Then both sides of (87) was operated with $d^{1/2} / dx^{1/2}$ to obtain

$$\frac{d^{1/2}}{dx^{1/2}} k = \sqrt{\pi} f(x) , \tag{88}$$

because these fractional operators (with suitable conditions on f) have the property that $D^{1/2} D^{-1/2} f = D^0 f$. Hence, when the fractional derivative of order ½ of constant k in (88) is calculated, $f(x)$ is determined. At this point of the text it is important to note that *the fractional derivative of a constant is not always equal to zero.*

Perhaps it was Fourier's integral formula and Abel's solution of the integral equation that drew Liouville's attention to do a major study of fractional calculus. The known result of derivatives of integral order

$$D^m e^{ax} = a^m e^{ax} \tag{89}$$

was the one with which Liouville began searching the fractional calculus and he extended this representation to derivatives of arbitrary order

$$D^v e^{ax} = a^v e^{ax} .$$

The arbitrary derivative of a function $f(x)$ that may be expanded in a series of the form

$$f(x) = \sum_{n=0}^{\infty} c_n e^{a_n x}, \mathrm{Re}(a_n) > 0 \tag{90}$$

is assumed to be

$$D^{v} f(x) = \sum_{n=0}^{\infty} c_n a_n^{v} e^{a_n x}, \tag{91}$$

which is Liouville's first formula for a fractional derivative. This has the drawback that it is applicable only to functions of the form given in (90).

Aware of this disadvantage, Liouville proposed the second definition, by starting with a definite integral related to gamma function

$$I = \int_0^{\infty} u^{a-1} e^{-xu} du, a > 0, x > 0$$

in which the change of variable $xu = t$ yields

$$x^{-a} = \frac{1}{\Gamma(a)} I.$$

Then Liouville invoked operations with D^{v} on both sides of this equation to obtain, with regard to (89), the following :

$$D^{v} x^{-a} = \frac{(-1)^{v}}{\Gamma(a)} \int_0^{\infty} u^{a+v-1} e^{-xu} du, \tag{92}$$

from which the second definition of fractional derivative is obtained as

$$D^{v} x^{-a} = \frac{(-1)^{v} \Gamma(a+v)}{\Gamma(a)} x^{-a-v}, a > 0. \tag{93}$$

But this too did not last long. The disadvantage was that it can only be useful to functions of the type $x^{-a} (a > 0)$.

William Center, in 1848, noticed that the fractional derivative of a constant, according to Lacroix-Peacock method, is not equal to zero (refer to such a mention above). Employing x^{0} to denote unity, Center finds the fractional derivative of unity of order ½, by letting $m = 0$ and $n = 1/2$ in the equation (83), is

$$\frac{d^{1/2}}{dx^{1/2}}x^0 = \frac{\Gamma(1)}{\Gamma\left(\frac{1}{2}\right)}x^{-1/2} = \frac{1}{\sqrt{\pi x}}. \tag{94}$$

Center points out, according to Liouville's "system" (refer to formula (93)), by letting $a = 0$ (even though Liouville assumed $a > 0$), the fractional derivative of unity equals zero because $\Gamma(0) = \infty$.

Bernhard Riemann developed his theory of fractional integration in his student days, but withheld the publication. It was published posthumously in his Gesammelte Werke [95]. He made a generalization of Taylor series and derived

$$D^{-v} f(x) = \frac{1}{\Gamma(v)} \int_c^x (x-t)^{v-1} f(t)dt + \Psi(x). \tag{95}$$

Riemann, due to some confusion in the lower limit c, added to his definition a complementary function $\Psi(x)$, which is essentially an attempt to provide a measure of the deviation from the law of exponents.This law, for example, is

$${}_cD_x^{-v}\, {}_cD_x^{-\mu} f(x) = {}_cD_x^{-\mu-v} f(x),$$

where the subscripts c and x on D refer to the limits of integration given in (95), and is true when lower limits are equal. In the present day definition, the formula (95) is free from $\Psi(x)$, see Cayley [14].

4.3.1 The Riemann-Liouville Definition

The paper of Sonin [105] is the earliest work that paved the way for, what is now called, the Riemann-Liouville definition. Letnikov [50] gave an extension of Sonin's paper, which had Cauchy's integral formula as the starting point. The nth derivative of Cauchy's integral formula is

$$D^n f(z) = \frac{n!}{2\pi i} \int_C \frac{f(\zeta)}{(\zeta - z)^{n+1}} ds. \tag{96}$$

Here generalization of $n!$ to arbitrary values raises no problem because $\nu ! = \Gamma(\nu+1)$. When n is not an integer, the integrand in (96) no longer contains a pole, but a branch point. Sonin and Letnikov did not mention , in their papers, the need of a branch cut for an appropriate contour.

The paper of Laurent [47] developed the theory of generalized operators and exposed that the theory of fractional calculus is closely connected with the theory of operators. He, like Sonin, also has began with Cauchy's integral formula, where the contour taken was open circuit on a Riemann surface. The contour integration resulted into the definition

$$ {}_0D_x^{-\upsilon} f(x) = \frac{1}{\Gamma(\upsilon)} \int_c^x (x-t)^{\upsilon-1} f(t)dt, \ \mathrm{Re}(\upsilon) > 0 \tag{97} $$

for integration to an arbitrary order.

For $x > c$ in (97), it is Riemann definition, given in (95), without a complementary function. Commonly used representation is when $c = 0$, that is

$$ {}_cD_x^{-\nu} f(x) = \frac{1}{\Gamma(\nu)} \int_0^x (x-t)^{\nu-1} f(t)dt, \mathrm{Re}(\nu) > 0. \tag{98} $$

This is often referred to as the Riemann-Liouville fractional integral formula.

$$ f(1/x) = O(x^{1-\epsilon}), \epsilon > 0 \tag{99} $$

is the sufficient condition that (98) converges. When c is negative infinity, (97) reduces to

$$ {}_{-\infty}D_x^{-\nu} f(x) = \frac{1}{\Gamma(\nu)} \int_{-\infty}^x (x-t)^{\nu-1} f(t)dt, \mathrm{Re}(\nu) > 0. \tag{100} $$

where

$$ f(-x) = O(x^{-\nu-\varepsilon}) \varepsilon > 0, x \to \infty \tag{101} $$

is the sufficient condition for the convergence of (100).

Commonly in use, in the recent years, is the Weyl fractional integral formula [113]

$$ {}_xW_{\infty}^{-v} f(x) = \frac{1}{\Gamma(v)} \int_x^{\infty} (t-x)^{v-1} f(t)dt, \mathrm{Re}(v) > 0, \tag{102}$$

in place of formula (100). It should be exclusively noted that the relation (97) is the Riemann version, whereas the formula (100) is called the Liouville version.

4.3.2 The Fractional Derivative

$D^n f(x)$, if n is a positive integer, will mean the n th derivative of $f(x)$, where $D = d/dx$ is the differentiation operator. In the event that n is not a positive integer, we notice that while we may assign a meaning to D^{-v} for $\mathrm{Re}(v) > 0$ (see Section 4.3.1), we are to assign the meaning to $D^v, \mathrm{Re}(v) > 0$.

Let $\mathrm{Re}(v) > 0$ and also let n is the smallest integer greater than $\mathrm{Re}(v)$, and let $v = n - v$. Then $0 < \mathrm{Re}(v) \le 1$. We difine the fractional derivative of $f(x)$ of order v as

$$ {}_cD_x^v f(x) = {}_cD_x^n \left[{}_cD_x^{-v} f(x) \right], x > 0. \tag{103}$$

Let us explain (103). Write it as

$$ {}_cD_x^v f(x) = {}_cD_x^n \left[\frac{1}{\Gamma(v)} \int_c^x (x-t)^{v-1} f(t)dt \right], x > 0. \tag{104}$$

Now, if $c = 0$ and $f(x) = \mu$, $\mu > -1$, then (104) gives

$$ {}_0D_x^{-v} x^{\mu} = \frac{1}{\Gamma(v)} \int_0^x (x-t)^{v-1} t^{\mu} dt,$$

which, for $\left(1 - \frac{t}{x}\right) = u$, reduces to

$$-\frac{1}{\Gamma(\nu)}\int_{1}^{0} x^{\nu-1}u^{\nu-1}[(1-u)x]^{\mu}\,xdu$$

i.e. $$=\frac{1}{\Gamma(\nu)}\int_{0}^{1} u^{\nu-1}(1-u)^{(\mu+1)-1}x^{\mu+\nu}du\,.$$

We obtain

$$_0D_x^{-\nu}x^{\mu}=\frac{\Gamma(\mu+1)}{\Gamma(\mu+\nu+1)}x^{\mu+\nu}, \tag{105}$$

provided that $\mathrm{Re}(\nu)>0, x>0$.

Besides Riemann Liouville definitions of fractional integral and differential operators, there are two additional definitions. One is due to Grünwald [38] and other is due to Marchaud [67].

Whatever that is mentioned above, is now presented in a very formal manner, by including some examples.

Definition 1 : Let f be a function of the class $\mathbb{C}$ and let $\mu>0$. Let m be the smallest integer that exceeds μ. Then the fractional derivative of f of order μ is defined by

$$D^{\mu}f(t)=D^{m}[D^{-\nu}f(t)], \mu>0, t>0 \tag{106}$$

if it exists, where $\nu=m-\mu>0$.

It is obvious that if μ is a positive integer, say p , then $D^{p}f(t)$ may exist for $t>0$ even if $f(t)$ is not of class $\mathbb{C}$, for instance $f(t)=t^{-1}$. But if f has p continuous derivatives on $(0,\infty)$, then (certainly) it is of class $\mathbb{C}$ and from

$$D^{p}f(t)=D^{p+1}\int_{0}^{t} f(\xi)d\xi=D^{p}f(t) \tag{107}$$

we notice that the relation (106) agrees with the definition of ordinary derivative.

Example (1) : Let $f(t)=t^{\lambda},\lambda>-1$. Let μ is a positive number and m is the smallest integer greater than μ. Then the fractional derivative of t^{λ} of order μ is, by (106),

$$D^{\mu}t^{\lambda}=D^{m}[D^{-v}t^{\lambda}],v=m-\mu>0. \tag{108}$$

But [by (105)]

$$D^{-v}t^{\lambda}=\frac{\Gamma(\lambda+1)}{\Gamma(\lambda+1+v)}t^{\lambda+v},t>0$$

and thus, from (108),

$$D^{\mu}t^{\lambda}=\frac{\Gamma(\lambda+1)}{\Gamma(\lambda+1+v)}D^{m}t^{\lambda+v}$$

i.e.

$$=\frac{\Gamma(\lambda+1)}{\Gamma(\lambda-\mu+1)}t^{\lambda-\mu},t>0. \tag{109}$$

Example (2) : Now let $f(t)=e^{at}$. Then from (106), we have

$$D^{\mu}e^{at}=D^{m}[D^{-v}e^{at}], \tag{110}$$

where μ,v and m have the same meaning, which are mentioned above. From the relation

$$D^{-v}e^{at}=E_{t}(v,a) \tag{111}$$

and employing

$$DE_{t}(v,a)=E_{t}(v-1,a), \tag{112}$$

we obtain

$$D^{m}E_{t}(v,a)=E_{t}(v-m,a)=E_{t}(-\mu,a),\mu=m-v \tag{113}$$

Hence, we conclude that

$$D^{\mu}e^{at}=E_{t}(-\mu,a),t>0. \tag{114}$$

Example (3) : We now obtain fractional derivative of $t^{\lambda}\log t,\lambda>-1$

From Definition 1, for $f(t)=t^{\lambda}\log t$,

$$D^{\mu}[t^{\lambda}\log t]=D^{m}[D^{-v}t^{\lambda}\log t], \tag{115}$$

$\mu, v,$ and m have usual meaning. We know [79] that the fractional integral of $t^{\lambda}\log t$ of order v, for $t>0$, is

$$D^{-v}[t^{\lambda}\log t]=\frac{\Gamma(\lambda+1)t^{\lambda+v}}{\Gamma(\lambda+v+1)}\left[\log t+\psi(\lambda+1)-\psi(\lambda+v+1)\right]. \tag{116}$$

Thus

$$\psi\equiv\psi(\lambda+1)-\psi(\lambda+v+1) \tag{117}$$

is the notation used for convenience for the following, obtained from (116)

$$D^{\mu}[t^{\lambda}\log t]=\frac{\Gamma(\lambda+1)}{\Gamma(\lambda+v+1)}D^{m}[t^{\lambda+v}(\log t+\psi)]. \tag{118}$$

Note that Ψ in (117) is a constant, independent of t.

Now by using Leibniz rule for the n th derivative of product of two functions, relation (115) can be written as

$$D^{\mu}[t^{\lambda}\log t]=\frac{\Gamma(\lambda+1)}{\Gamma(\lambda+v+1)}\sum_{k=0}^{m}\binom{m}{k}$$

$$\times[D^{m-k}t^{\lambda+v}][D^{k}(\log t+\Psi)]. \tag{119}$$

Now,

$$D^{m-k}t^{\lambda+v}=\frac{\Gamma(\lambda+v+1)}{\Gamma(\lambda+v-m+k+1)}t^{\lambda+v-m+k},$$

for $k=0,1,2,\ldots,n$, whereas

$$D^{\circ}[\log t+\Psi]=\log t+\Psi$$

and

$$D^{k}[\log t+\Psi]=(-1)^{k-1}(k-1)!t^{-k}, k=1,2,\ldots,m.$$

Hence (119) becomes

$$D^{\mu}[t^{\lambda}\log t]=\frac{\Gamma(\lambda+1)}{\Gamma(\lambda-\mu+1)}t^{\lambda-\mu}$$

$$\times\left[\ [\log t+\psi(\lambda+1)-\psi(\lambda-\mu+m+1)\right.$$

$$\left.-\sum_{k-1}^{m}\frac{(-1)^k m!\Gamma(\lambda-\mu+1)}{k(m-k)!\Gamma(\lambda-\mu+k+1)}\right],\quad v=m-\mu. \tag{120}$$

With little more calculations [79], the required fractional derivative is

$$D^{\mu}[t^{\lambda}\log t]=\frac{\Gamma(\lambda+1)}{\Gamma(\lambda-\mu+1)}t^{\lambda-\mu}$$

$$\times\left[\log t+\psi(\lambda+1)-\psi(\lambda-\mu+1)\ \right]. \tag{121}$$

4.3.3 Examples

Exclusively, the relation (98) suggests the following definition : Let $\mathrm{Re}(\nu)>0$ and $f(x)$ is piecewise continuous on $(0,\infty)$ and integrable on any finite subinterval $(0,\infty)$. Then for $t>0$,

$${}_0D_t^{-\nu}f(t)=\frac{1}{\Gamma(\nu)}\int_0^t(t-\xi)^{\nu-1}f(\xi)d\xi, \tag{122}$$

which is called the Riemann-Liouville fractional integral of f of order ν. In what follows we illustrate (122).

Example (1) : Let $f(t)=e^{at}, e^{at}\in\mathbb{C}$. Then from (122), we have

$${}_0D_t^{-\upsilon}e^{at}=\frac{1}{\Gamma(\upsilon)}\int_0^t(t-\xi)^{\upsilon-1}e^{a\xi}d\xi,\mathrm{Re}(\upsilon)>0.$$

Using transformation $t-\xi=x$, right-hand side is

$$-\frac{1}{\Gamma(\upsilon)}\int_t^0 x^{\upsilon-1}e^{a(t-x)}dx$$

$$=\frac{e^{at}}{\Gamma(\upsilon)}\int_0^t x^{\upsilon-1}e^{-ax}dx,x>0.$$

This is closely related to the incomplete gamma function

$$\gamma^*(\nu,z)=\frac{z^{-\nu}}{\Gamma(\nu)}\int_0^z t^{\nu-1}e^{-t}dt,\mathrm{Re}(z)>0 \tag{123}$$

which reduces the right-hand side to

$$t^{v}e^{at}\gamma^{*}(v,at). \tag{124}$$

Solution is completed.

Example (2) : Let $f(t)=\cos at$ in (122). Then we have

$$_{0}D_{t}^{-v}\cos at=\frac{1}{\Gamma(v)}\int_{0}^{t}(t-\xi)^{v-1}\cos a(t-\xi)d\xi, v>0.$$

Set $t-\xi=x,$ we get

$$=\frac{1}{2\Gamma(v)}\int_{0}^{t}x^{v-1}\{e^{iax}+e^{-iax}\}dx$$

$$=\frac{1}{2}\left[\frac{1}{\Gamma(v)}\int_{0}^{t}x^{v-1}e^{iax}dx+\frac{1}{\Gamma(v)}\int_{0}^{t}x^{v-1}e^{-iax}dx\right].$$

Using (123) and invoking the formula

$$E_{z}(v,a)=z^{v}e^{az}\gamma^{*}(v,az), \tag{125}$$

we obtain

$$_{0}D_{t}^{-v}\cos at=\frac{1}{2}[E_{z}(v,ia)+E_{z}(v,-ia)],$$

which, further, reduces to

$$_{0}D_{t}^{-v}\cos at=C_{z}(v,a), \tag{126}$$

by using

$$C_{z}(v,a)=\frac{1}{2}[E_{z}(v,ia)+E_{z}(v,-ia)]. \tag{127}$$

Similarly, solution for $f(t)=\sin at$ can be obtained.

Example (3) : In the formula (124), set $v=1/2,$ we get

$$_{o}D_{t}^{-1/2}e^{at}=e^{at}t^{1/2}\gamma^{*}\left(\frac{1}{2},at\right)$$

$$=E_{t}\left(\frac{1}{2},a\right) \tag{128}$$

Now use error function Erf(.) in terms of incomplete gamma function, that is,

$$Erf\ \ x = x\gamma^{*}\left(\frac{1}{2}, x^{2}\right), \tag{129}$$

to obtain (129) as

$${}_{o}D_{t}^{-1/2}e^{at} = t^{1/2}e^{at}\frac{1}{\sqrt{at}}Erf\left(\sqrt{at}\right),$$

i.e.

$${}_{o}D_{t}^{-1/2}e^{at} = \frac{e^{at}}{\sqrt{a}}Erf\left(\sqrt{at}\right). \tag{130}$$

Example (4) : If $f(t) = \cos^{2} at$ in (122), then

$${}_{0}D_{t}^{-v}\cos^{2} at = \frac{1}{\Gamma(v)}\int_{0}^{t}(t-\xi)^{v-1}\cos^{2} a(t-\xi)d\xi$$

$$= -\frac{1}{\Gamma(\upsilon)}\int_{t}^{0}u^{\upsilon-1}\cos^{2} au, \quad t-\xi = u$$

i.e. $${}_{0}D_{t}^{-\upsilon}\cos^{2} at = \frac{1}{2\Gamma(\upsilon)}\left[\int_{0}^{t}u^{\upsilon-1}du + \int_{0}^{t}u^{\upsilon-1}\cos 2au\ du\right].$$

This yields the solution

$${}_{0}D_{t}^{-v}\cos^{2} at = \frac{t^{v}}{2\Gamma(v+1)} + \frac{1}{2}C_{t}(v,2a). \tag{131}$$

Example (5) : If we take $f(t) = (a-t)^{\lambda}, f \in \mathbb{C},\ a > t > 0,$ then from (122), we have

$${}_{0}D_{t}^{-v}(a-t)^{\lambda} = \frac{1}{\Gamma(v)}\int_{0}^{t}(t-\xi)^{v-1}(a-\xi)^{\lambda}d\xi. \tag{132}$$

Now set $$x = \frac{t-\xi}{a-\xi},$$

a bilinear transformation, which transforms (132) to

$$ {}_0D_t^{-\nu}(a-t)^{\lambda} = \frac{(a-t)^{\nu+\lambda}}{\Gamma(\nu)} \int_0^{t/a} x^{\nu-1}(1-x)^{-\nu-\lambda-1}\,dx. \tag{133} $$

Invokinig the incomplete Beta function

$$ B_{\tau}(x,y) = \int_0^{\tau} t^{x-1}(1-t)^{y-1}\,dt,\ \mathrm{Re}(x) > 0,\ 0 < \tau < 1 \tag{134} $$

the relation (133) yields the solution

$$ {}_0D_t^{-\nu}(a-t)^{\lambda} = \frac{(a-t)^{\nu+\lambda}}{\Gamma(\nu)} B_{t/a}(\nu, -\nu-\lambda). \tag{135} $$

Chapter 5
Distributions, Boehmians and Integral Transforms

5.1 Kontorovich-Lebedev Transformation of Tempered Distribution and Tempered Beohmains [54]

Section (A) : This section gives investigation for the Kontorvich-Lebedev transform of testing functions of rapid descent, and of tempered Boehmians. The Kontorovich-Lebedev transform (KLT) arises naturally by the method of separation of variables to derive solutions of boundary value problems, which are formulated in terms of cylindrical coordinates.

The KLT of a function *f*(*x*), defined on the positive real line, is [cf. [103 ;p.359]]

$$\tilde{f}(\tau) = \int_0^\infty x^{-1} f(x) K_{i\tau}(x) dx. \tag{136}$$

Its inversion formula and Parseval relation, respectively, are

$$f(x) = \frac{2}{\pi^2} \int_0^\infty K_{i\tau}(x) \tau \sinh(\pi\tau) \tilde{f}(\tau) d\tau, \tag{137}$$

and

$$\int_0^\infty x^{-1} f(x) g(x) dx = 2/\pi^2 \int_0^\infty \tau \sinh(\pi\tau) \tilde{f}(\tau) \tilde{g}(\tau) d\tau, \tag{138}$$

where $K_{i\tau}(x)$ is the Macdonald function.

The extension of KLT for distributions is

$$\tilde{f}(\tau) = \int_0^\infty x^{-1/2} f(x) K_{i\tau}(x) dx \tag{139}$$

and the corresponding inversion formula is

$$f(x) = \frac{2}{\pi^2} \int_0^\infty \tau \sinh(\pi\tau) x^{-1/2} \tau K_{i\tau}(x) G(\tau) d\tau, \tag{140}$$

x and τ are positive real vaviables.

The space $k'_\alpha(I)$ is the dual of $k_\alpha(I)$, where the latter is the collection of infinitely differentiable complex-valued functions $\varphi(x)$ on $I = (0, \infty)$, having the norm

$$\gamma(\varphi(t)) = \sup_{0<t<\infty} \left| \xi(t) \Delta_t^k \varphi(t) \right| < \infty, \tag{141}$$

for $k = 0,1,2,...$ and $\Delta_t = t^2 D^2 + 2tD - t^2, D = d/dt$,

and

$$\xi(t) = \begin{cases} t^{1/2} & ,0<t<1 \\ t^{\frac{3}{2}-\alpha} \log t & ,t>2. \end{cases} \tag{142}$$

Since $F(\tau)$, the distributional KLT, is a smooth function for $\tau > 0$ and $F \in k'_\alpha(I)$, its *mth* derivative is

$$D_x^m F(\tau) = \left\langle f(x), D_x^m x^{-1/2} K_{i\tau}(x) \right\rangle, \tag{143}$$

estimation of which is

$$F(\tau) = \begin{cases} O(1) & ,t \to 0_+ \\ O(x^{2\gamma-\alpha+1} e^{-x\pi/\gamma} \log x) & ,\tau \to \infty \end{cases} \tag{144}$$

where $\alpha > 0$ and γ is a non-negative integer depending on *f*.

In what follows, is the theorem which determines the KLT of testing function of rapid descent. Zemanian [115] defined $\varphi(t)$ in the test function space *S*, whenever

$$|t|^m \left| \frac{\partial^{k_1+k_2....+k_n}}{\partial t_1^{k_1}, \partial t_2^{k_2} ... \partial t_n^{k_n}} \varphi(t_1....t_n \right| \le C_{mk_1....k_n} \tag{145}$$

over the entire R^N and $\varphi(t)$ is a complex-valued infinitely differentiable function for $|t| \to \infty$. The partial derivatives decrease to zero faster than every power of $1/|t|$. We write (145) as

$$|t|^m \left|D^k \varphi(t)\right| \le C_{mk}. \tag{146}$$

Elements of S are called the testing functions of rapid descent.

Theorem 1: If $f(t)$ is in the space S, then its KLT, given by (139), is also in S.

Proof : Since $f(t) \in S$, its KLT exists. Its *mth* derivative (141), by using (143) and (144), can be expressed as

$$\left|F^m(\tau)\right| \le C, \tau > 0. \tag{147}$$

Indeed $F^m(\tau) \to 0$ as $t \to \infty$.

The theorem is, thus, proved.

Corollary 1: Every function in S_τ is the KLT of a function in S_t. Indeed, the inverse provides a one-to-one correspondence between S_τ and S_t.

Proof: We invoke the proof of the above theorem to the inverse KLT, that is given by (140), of an arbitrary function $F(\tau)$ in S_τ . We assert, at once, $f(x) \in S_\tau$.

Theorem 2: The KLT and its inverse are continuous linear mapping of S onto itself.

Proof: We know that the KLT is linear (see [103]) and the sequence $\{f_v(t)\}_{v=1}^{\infty}$ converges in S to zero, owing to

(i) $f_v(t)$ is in S

(ii) $|t|^m \left|D^k f_v(t)\right| \le C_{mk},$

where m and k are non- negative integers, and

(iii) $\{D^K \varphi_v(t)\}_{v=1}^{\infty}$ converges uniformly over every bounded domain R^N (see equations (145) and (146)).

Moverover, relation (147) suggests the convergence in *S*.

Section (B) : We, in this part, deal with the KLT of tempered distributions. The distribution f is said to be of slow growth (or temperate) if it is a continuous linear functional on the test function space S of rapid descent or tempered distribution, subject to the conditions of linearity and continuity. Such space (of slow growth) is denoted by S' [see[115]].

In the context of [115], we observe that the Fourier transform of distributions of slow growth is defined by virtue of the Parseval relation. Similarly, if the locally integrable function $f(t)$ is absolutely intagrable for $0<t<\infty, \varphi$ is a test function of rapid descent, and their KLT transform $\tilde{f}$ and $\tilde{\varphi}$ exist and (at the same time) satisfy

$$\langle f(x), \varphi(x) \rangle = \langle \tilde{f}(t), \tilde{\varphi}(t) \rangle, \tag{148}$$

which is the Parseval relation (138) for the distribution space, then $\tilde{f}$ is a functional on S as φ traverses S. In other words, the KLT $\tilde{f}$ of a distribution f of slow growth is defined as the functional, which assigns to each φ in S, the contidions similar to that f assigns to the KLT.

Theorem 3 : If f is a distribution of slow growth, then its KLT $\tilde{f}$ is a distribution of slow growth too.

Proof : By virtue of the continuity, $\tilde{f}$ is linear.

If the sequence $\{\varphi_v\}_{v=1}^{\infty}$ converges in *S* to zero, then (indeed) the sequence $\{\tilde{\varphi}_v\}_{v=1}^{\infty}$ converges also in *S* to zero. Now, as $v \to \infty$,

$$\langle \tilde{f}, \tilde{\varphi}_v \rangle = \langle f, \varphi_v \rangle \to 0, \tag{149}$$

and hence, $\tilde{f}$ is a continuous linear functional on *S*.

Further, since the Parseval relation serves as the inverse of the KLT of an arbitrary distribution f in S', the KLT and its inverse provide one-to-one mapping of S' onto itself and (also) they are continuous.

Section (C) : This part investigates the KLT of tempered Boehmians. The quotient of sequence f_n/φ_n, numerator of which belongs to some set G and the denominator is a delta sequence, permits us to write the convolution

$$f_m * \varphi_n = f_n * \varphi_m, \forall m,n \in N. \tag{150}$$

The quotient of sequences f_m/φ_m and g_n/ψ_n are said to be equivalent if

$$f_m * \psi_n = g_n * \varphi_m, \forall m,n \in N. \tag{151}$$

The equivalence classes are called Boehmians, space of all Boehmians is denoted by β.

A complex-valued function f on R^N is said to be slowly increasing if there exists a polynomial p on R^N such the $f(x)/p(s)$ is bounded. The space of all slowly increasing continuous functions on R^N is denoted by ϑ. If $f_n \in \vartheta$, then $\{\varphi_v\}_{v=1}^{\infty}$ is a delta sequence under such conditions those are usual. The space of equivalence class of quotients of sequences is a denoted by β_ϑ, and its elements are called tempered Boehmians.

If $F = [f_n/\varphi_n] \in \beta_\vartheta$ and $f_n \in S$ for all $n \in N$, then f is called rapidly decreasing Boehmian, space of which is denoted by β_S. It may be noted that the KLT of tempered Boehmian, and $\tilde{f}(\tau)$ of slowly increasing function f is the Schwartz distribution, see equation (148). The tempered Boehmian $F = [f_n/\varphi_n]$ of the KLT $\tilde{F}$, is the limit of the sequence $\{\tilde{f}_n\}_{n=1}^{\infty}$ in $\mathcal{D}'$. Also, that f/φ and $\dfrac{f}{\varphi}$ the denote, respectively, the convolution quotient and the usual quotient.

Theorem 4: It $[f_n/\varphi_n]=\beta_\vartheta$, then the sequence $[\tilde{f}_n]$ converges in $\mathcal{D}'$. Moreover, if $[f_n/\varphi_n]=[g_n/\gamma_n]\in\beta_\vartheta$, then sequences $\{\tilde{f}_n\}$ and $\{\tilde{g}_n\}$ converge to the same limit for the KLT of tempered Boehmians.

The details of proof may be seen in Banerji and Loonker [4].

Theorem 5 : Let $F=[f_n/\varphi_n]$ and $G=[g_n/\gamma_n]$.

Then

(I) $\tilde{G}$ is an infinitely differentiable function

(II) $\tilde{f}*\tilde{g}=f.g.$

Proof (I) : Let $G=[g_n/\gamma_n]\in\beta_S$ and let U is a bounded open subset of R^N. Then there exists $m\in N$ such that $\gamma_m>0$ on U. We now have

$$\tilde{G}=\lim_{n\to\infty}\tilde{g}_n=\lim_{n\to\infty}\frac{\tilde{g}_n\tilde{\gamma}_m}{\tilde{\gamma}_m}$$

$$=\lim_{n\to\infty}\frac{\tilde{g}_m\tilde{\gamma}_n}{\tilde{\gamma}_m}$$

$$=\frac{\tilde{g}_m}{\gamma_m}\lim_{n\to\infty}\tilde{\gamma}_n$$

i.e. $$\tilde{G}=\frac{\tilde{g}_m}{\tilde{\gamma}_m}\text{ on }U.$$

Further, since $\tilde{g}_m.\tilde{\gamma}_m\in S$ and $\tilde{\gamma}_m>0$ on U, $\tilde{G}$ is an infinitely differentiable function on U.

(II) : Let $F=[f_n/\varphi_n]\in\beta_\vartheta$ and $G=[g_n/\gamma_n]\in\beta_S$. If $\varphi\in\mathcal{D}$. Then there exists $m\in N$ such that $\tilde{\gamma}_m>0$ on the support of φ.

We know [cf.[114], p,213] that the convolution for the KLT, defined as here, on $(0,\infty)$

$$(f*g)(x)=\frac{1}{2\sqrt{x}}\int_0^\infty\int_0^\infty\exp\left(\frac{-xu}{2y}-\frac{xy}{2u}-\frac{uy}{2x}\right)f(u)g(y)\frac{dudy}{\sqrt{uy}}\tag{153}$$

exists for almost all $x \in (0,\infty)$ for functions f and g, respectively, that is (in other words)

$$\tilde{f} * \tilde{g} = \tilde{f}.\tilde{g}\,. \tag{154}$$

From equations (148),(154) and the Theorem 4, we obtain

$$\left(\tilde{F} * \tilde{G}\right)(\varphi) = \lim_{n\to\infty} \left(\tilde{f}_n * \tilde{g}_n\right)(\varphi)$$

$$= \lim_{n\to\infty} \left(\tilde{f}_n, \tilde{g}_n \varphi\right)$$

i.e.
$$= \lim_{n\to\infty} \tilde{f}_n \left(\frac{\tilde{g}_m \tilde{\gamma}_n \varphi}{\tilde{\gamma}_m} \right)$$

i.e.
$$= \lim_{n\to\infty} \tilde{f}_n \left(\tilde{G} \varphi \tilde{\gamma}_n\right),$$

which when simplied, yields

$$\left(\tilde{F} * \tilde{G}\right)(\varphi) = \left(\tilde{F}\,\tilde{G}\right)\varphi\,.$$

Thus the proof is completed.

5.2 Cauchy Representation of Integrable and Tempered Boehmians [55]

A relation between the Cauchy representation of the Fourier transform of the functions in L_2 − spaces and a decomposition of the Fourier transform into two parts, each of which gives an analytic function in a half plane, defines that the decomposed transform is convergent for classes of functions larger than those in L_2 − space. We intend to discuss the Cauchy representation of integrable Boehmians by using a relation between the Cauchy representation and the Fourier transform, and the properties of the former in L_1 − space.

5.2.1 Definitions and Terminologies

Property 1 : Let $f \in L_2, g(w) = F(f,w)$ and $g(z)$ is a Cauchy representation of $g(z)$, where $z{=}x{+}iy$. Then

$$\hat{g}(z) = \begin{cases} \int_0^{\infty} f(t)e^{itz}dt, y > 0 \\ -\int_{-\infty}^{0} f(t)e^{itz}dt, y < 0. \end{cases} \tag{155}$$

Equation (155) holds even if the assumption $f \in L_2$ is replaced by $f, g \in L_1$.

Property 2: Let $g \in L_2, f(t) = F^{-1}(g,t)$ and $f(z), z = x + iy,$ be the Cauchy representation of *f*. Then

$$\hat{f}(z) = \begin{cases} (2\pi)^{-1} \int_0^{\infty} g(w)e^{-iwz}dw, y > 0 \\ -(2\pi)^{-1} \int_{-\infty}^{0} g(w)e^{-iwz}dw, y < 0. \end{cases} \tag{156}$$

Definition 1 : A function *f* is called a tempered function (or a function of slow growth, or a slowly increasing function) if *f* is continuous and for some α,

$$|f| = 0\left((|t|)^{a}\right). \tag{157}$$

Definition 2 : The generalized Fourier transform and the generalized inverse Fourier transform of a tempered function *f* are defined by

$$\hat{F}(f,z) = \begin{cases} \int_0^{\infty} f(t)e^{itz}dt, y > 0 \\ -\int_{-\infty}^{0} f(t)e^{itz}dt, y < 0. \end{cases} \tag{158}$$

and

$$\hat{F}^{-1}(z) = \begin{cases} (2\pi)^{-1} \int_{-\infty}^{0} f(t)e^{-itz}dt, y > 0 \\ -(2\pi)^{-1} \int_0^{\infty} f(t)e^{-itz}dt, y < 0. \end{cases} \tag{159}$$

Property 3 : $\hat{F}(f,z)$ and $\hat{F}^{-1}(f,z)$, defined in equations (158) and (159), respectively, are analytic for $y \neq 0$.

Following identities hold true between $\hat{F}(f,z)$ and the classical Fourier transform and

$$\left.\begin{aligned}\hat{F}(f,x+i\varepsilon = F(f(t)H(t)e^{-\varepsilon t},x),\varepsilon>0\\ \hat{F}(f,x-i\varepsilon)=-F(f(t)H(-t)e^{-\varepsilon t},x),\varepsilon>0\end{aligned}\right\}, \tag{160}$$

where $H(t)$ is a positive function, convenient for analysis of result of inverse theorem, and has a positive Fourier transform whose integral is easily calculated. Thus, we have

$$\hat{F}(f,x-i\varepsilon)=\hat{F}(f,x-i\varepsilon)=F\left(e^{-\varepsilon|t|}f(t),\varepsilon\right)$$

$$=\int_{-\infty}^{\infty}e^{-\varepsilon|t|}f(t)e^{itx}dt. \tag{161}$$

Property 4 : Let f is a tempered function. Then the generalized Fourier transform of f is expressed as

$$\langle F(f),\varphi\rangle=\lim_{\varepsilon\to 0}\int_{-\infty}^{\infty}\left(\hat{F}(f,x+i\varepsilon)-\hat{F}(F,x-i\varepsilon)\right)\varphi(x)dx \tag{162}$$

for all $\varphi\in S$ and $\varepsilon>0$.

Property 5 : For a given tempered function f, the inversion formula for the generalized Fourier transform $\hat{F}(f,z)$ has the property

$$(2\pi)^{-1}\int_{-\infty}^{\infty}\left(\hat{F}(f,x+i\varepsilon)-\hat{F}(f,x-i\varepsilon)\right)\; e^{-ixt}dt=e^{-\varepsilon|t|}f(t),\varepsilon>0 \tag{163}$$

and

$$\lim_{\varepsilon\to 0}F^{-1}\;\hat{F}(f,x+i\varepsilon)-\hat{F}(f,x-i\varepsilon)\;,t)=f(t),\varepsilon>0. \tag{164}$$

Property 6 : Let T is a functional for the *mth* derivative of the tempered function defined in L_1. Then the generalized Fourier transform of T is defined by

$$\hat{F}(T,z)=(-iz)^m\hat{F}(f,z). \tag{165}$$

Property 7 : Let $T\in S'$ and $\hat{F}(T,z)$ is a generalized Fourier transform of T. Then $\hat{F}(t,z)$ is an analytic (Cauchy) representation of $\hat{F}(T)$ in the sense that

$$\lim_{\varepsilon\to 0}\left\langle\hat{F}(f,x+i\varepsilon)-\hat{F}(f,x-i\varepsilon),\varphi(x)\right\rangle=\langle F(T),\varphi\rangle,\varphi\in S, \tag{166}$$

where S' is the space of linear functional on S, or the space of tempered distributions. Other properties and definitions may be seen in [55].

5.2.2 Cauchy Representation for Integrable Boehmians :

In the analysis of the problem in this section, we need following properties:

Property 8 : Suppose $f \in L_1, F(f) \in L_1$ and $F = F(f)$. Then the Cauchy representation of $\hat{F}(z)$, by means of the Cauchy kernel, is given by

$$\hat{F}(z) = \hat{F}(f, z), \tag{167}$$

$$\hat{f}(z) = \hat{F}(f, z), z = x + iy. \tag{168}$$

As a consequence of this property, if $f \in S$ and F denote $F(f)$, then

$$\hat{F}(z) = \hat{F}(f, z), y_1, y_2, \ldots, y_n \neq 0. \tag{169}$$

Property 9 : If $f, g \in L_1$ and $h(x) = (f * g)(x)$, then

$$\hat{h}(z) = \int_{-\infty}^{\infty} f(t)\hat{g}(z-t)dt \tag{170}$$

$$= \int_{-\infty}^{\infty} g(t)\hat{f}(z-t)dt. \tag{171}$$

This property is called the Cauchy representation of convolutions of functions belonging to L_1 – space.

Considering L_1 as the space of complex-valued Lebesgue integrable function on the real line R, the norm of a function f in L_1 is given by

$$\|f\| = \int_R |f(x)|dx. \tag{172}$$

If $f, g \in L_1$, then the convolution $f * g$, given by

$$f * g(x) = \int_R f(u)g(x-u)du \tag{173}$$

is an element of the space L_1, and

$$\|f * g\| \leq \|f\| \cdot \|g\| \tag{174}$$

Lemma 1 : If $[f_n / \delta_n] \in B_{L_1}$, then the sequence

$$\hat{f}_n(z) = \int_R f(t)e^{itz}(\sigma H(\sigma t))dt$$

converges uniformly on each compact set in *R*.

Proof : If (δ_n) represents a delta sequence, then (δ_n) converges on each compact set to the constant function 1. Hence, for each compact set $K, \hat{\delta}_K > 0$ on *K* for all most all $k \in K$ and

$$\hat{f}_n = \hat{f}_n \frac{\hat{\delta}_k}{\hat{\delta}_k} = \frac{(f_n + \delta_k)^\wedge}{\hat{\delta}_k} = \frac{(f_k + \delta_n)^\wedge}{\hat{\delta}_k} = \frac{\hat{f}_k}{\hat{\delta}_k}\hat{\delta}_n \quad \text{on } K.$$

The Cauchy representation of an integrable Boehmain $F = [f_n / \delta_n]$, indeed, can be defined as the limit of the sequence $(\hat{f}_n)$ in the space of continuous function on *R*. This proves that the Cauchy representation of an integrable Boehmian is a continuous function. This completes the proof of the lemma.

Theorem 6 : Let $F, G \in B_{L_1}$. Then

(i) $(\lambda F)^\wedge = \lambda \hat{F}$ (λ is any complex number), and

$(F + G)^\wedge = \hat{F} + \hat{G}$

(ii) $(F * G)^\wedge = \hat{F}\hat{G}$

(iii) $\hat{F}(f^{(m)}, z) = (-iz)^m \hat{F}(f, z)$

(iv) if $\hat{F} = 0$, then $F = 0$

if $\Delta - \lim_{n\to\infty} F_n = F$, then $\hat{F}_n \to F$ uniformly on each compact set.

Proof : Appealing to Properties 8 and 9 (Equations (167)-(171)), that is the relation between Cauchy representation and the Fourier transform, proofs of (i)-(ii) are natural. Proof of (iv) can be completed by employing uniqueness theorem of Fourier transform (if $F \in L_1$ and $\hat{f}(t) = 0$ for all $t \in R^1$, then $f(x) = 0$ almost everywhere in L_1. Now, we prove part (v):

We have

$$\delta - \lim_{n\to\infty} F_n - F \Rightarrow \hat{F}_n \to \hat{F}$$

uniformly on each compact set. Let (δ_n) be a delta sequence such that

$$F_n * \delta_k, F * \delta_k \in L_1, \forall n, k \in N$$

and

$$\|(F_n - F) * \delta_k\| \to 0 \text{ as } n \to \infty, \text{ for } k \in N.$$

Let K be a compact set in R. Then $\hat{\delta}_k > 0$ on K for some $k \in N$. Since δ_k is a continuous function, we observe

$$\hat{F}_n . \hat{\delta}_k \to \hat{F}\hat{\delta}_k$$

uniformly on K. Since

$$\hat{F}_n . \hat{\delta}_n - \hat{F} . \delta_k = ((F_n - F) * \delta_k)^{\wedge}$$

and

$$\|(F_n - F) * \delta_k\| \to 0, \text{ as } n \to \infty,$$

part (v) is proved and also the theorem is completely proved.

Theorem 7: Let $f \in B_{L_1}$ and

$$f_n(t) = \frac{1}{2\pi} \int_{-\infty}^{\infty} \hat{F}(z) e^{itz} dt. \tag{175}$$

Then $\delta - \lim_{n\to\infty} f_n = F$ and hence, also

$$\Delta - \lim_{n\to\infty} f_n = F.$$

Proof : Let $F = [g_n / \delta_n,], k \in N$. Then

$$(f_n * \delta_k)(t) = \int_R f_n(t-u) \delta_k(u) du$$

i.e.
$$= \frac{1}{2\pi} \int_R \int_{-\infty}^{\infty} e^{itz} \hat{F}(z) \delta_k(u) dt du$$

$$= \frac{1}{2\pi} \int_{-\infty}^{\infty} e^{itz} \hat{F}(z) \hat{\delta}_k(t) dt$$

i.e.
$$(f_n * \delta_k)(t) = \frac{1}{2\pi} \int_{-\infty}^{\infty} e^{itz} F * \delta_k(t) dt.$$

By Lemma 1, $\|f_n * \delta_k - F * \delta_k\| \to 0$ as $n \to \infty$. Since k is an arbitrary positive integer, $\delta - \lim_{n\to\infty} f_n = F$. Proof of the theorem is completed.

5.2.3 Cauchy Representation for Tempered Boehmians

Some tempered Boehmians are natural consequences of tempered distributions. Using

$$\hat{F}(t,z) = \int_{E^n} f(t)e^{itz}(\sigma H(\sigma t))dt, \tag{176}$$

we investigate the Cauchy representation of tempered Boehmians. Terminologies used are those of Mikusinski [78]. Let J be the space of slowly increasing infinitely differentiable complex-valued functions on R^N (f is called slowly increasing if there exists a polynomial p such that $|f(x)| \le p(x), \forall x \in R^N$). $\mathcal{D}$ is the space of all infinitely differentiable complex-valued functions on R^N with compact support.

A pair of sequence (f_n, φ_n) is called a quotient of sequences f_n / φ_n if $f_n \in J, \forall n \in N$ and $\{\varphi_n\}$ is a delta sequence. If the function space is of slowly increasing infinitely differentiable complex-valued functions, then the space of Boehmians, that we discuss in this section, consists of tempered Boehmians, which is denoted by B_J.

Theorem 8 : If $[f_n / \varphi_n] \in B_J$, then the sequence $\{\hat{f}_n\}$ converges in $\mathcal{D}'$. Moreover, if $[f_n / \varphi_n] = [g_n / \gamma_n] \in B_J$, then the sequences $\{\hat{f}_n\}$ and $\{\hat{g}_n\}$ converge to same limit.

Proof : Let $\varphi \in \mathcal{D}$ and the $k \in N$ be such that $\hat{\varphi} > 0$ on the support of φ. Since

$$f_n * \varphi_m = f_m * \varphi_n, \quad \forall m, n \in N$$

and

$$\hat{f}_n * \hat{\varphi}_m = \hat{f}_m * \hat{\varphi}_n,$$

$$\hat{f}_n(\varphi) = \hat{f}_n\left(\frac{\varphi\hat{\varphi}_k}{\hat{\varphi}_k}\right)$$

i.e. $$= \left(\hat{f}_n\hat{\varphi}_k\right)\left(\frac{\varphi}{\hat{\varphi}_k}\right)$$

i.e. $$= \left(\hat{f}_k\right)\left(\frac{\varphi\hat{\varphi}_n}{\hat{\varphi}_k}\right).$$

Since the sequence $[\varphi\hat{\varphi}_n / \hat{\varphi}_k]$ converges to $\frac{\varphi}{\hat{\varphi}_k}$ in $\mathcal{D}$, the sequence $\left\{\left(\hat{f}_n, \varphi\right)\right\}$ converges in $\mathcal{D}$. This proves that $\left\{\hat{f}_n\right\}$ converges in $\mathcal{D}'$. Further, assuming $[f_n / \varphi_n] = [g_n / \gamma_n] \in B_J$, and

$$h_n = \begin{cases} f_{\frac{n+1}{2}} * \gamma_{\frac{n+1}{2}}, & n \text{ is odd} \\ g_{\frac{n}{2}} * \varphi_{\frac{n}{2}}, & n \text{ is even} \end{cases}$$

and

$$\delta_n = \begin{cases} \varphi_{\frac{n+1}{2}} * \gamma_{\frac{n+1}{2}}, & n \text{ is odd} \\ \varphi_{\frac{n}{2}} * \gamma_{\frac{n}{2}}, & n \text{ is even} \end{cases}$$

we find $[h_n / \delta_n] = [f_n / \varphi_n] = [g_n / \gamma_n]$. Hence, the sequence $\{\hat{h}_n\}$ converges in $\mathcal{D}'$. Moreover,

$$\lim_{n\to\infty} \hat{h}_{2n-1}(\varphi) = \lim_{n\to\infty} (f_n * \gamma_n)^\wedge(\varphi)$$

$$= \lim_{n\to\infty} \left(\hat{f}_n\hat{\gamma}_n\right)(\varphi) = \lim_{n\to\infty} (\hat{f})(\hat{\gamma}\varphi)$$

$$= \lim_{n\to\infty} \hat{f}_n(\varphi).$$

Hence, we have proved that $\{\hat{h}_k\}$ and $\{\hat{f}_n\}$ have the same limit. Similarly, we can prove that $\{\hat{h}_n\}$ and $\{\hat{g}_n\}$ will possess same limit. The therorem is, thus, completely proved.

Theorem 9 : Let $F=[f_n/\varphi_n]\in B_J$ and $G=[g_n/\gamma_n]\in B_S$.

Then

(i) $(\partial F/\partial x_m)^\wedge = ix_m\hat{F}$

(ii) $\hat{G}$ is an infinitely differentiable function

(iii) $(F*G)^\wedge = \hat{F}\hat{G}$, and

(iv) $\hat{F}\hat{\varphi}_n = \hat{f}_n, \forall n\in N.$

Proof: (i) Left hand side is expressed as

$$\left(\frac{\partial F}{\partial x_m}\right)^\wedge = \left[\left(\frac{\partial f_n}{\partial x_m}*\varphi_n\right)/(\varphi_n*\varphi_n)\right]^\wedge$$

$$=\lim_{n\to\infty}\left(\frac{\partial f_n}{\partial x_m}*\varphi_n\right)^\wedge = \lim_{n\to\infty} ix_m\hat{f}\hat{\varphi}_m = ix_m\hat{F}. \tag{177}$$

Now the proof can be completed by virtue of Theorem 8 and invoking $[f_n/\varphi_n]=[(f_n*\varphi_n)/(\varphi_n*\varphi_n)]$.

(ii) Let $G=[g_n/\gamma_n]\in B_S$ and U be a bounded open subset of R^N.

Then there exists $m\in N$ such that $\hat{\gamma}_m>0$ on U

$$\hat{G}=\lim_{n\to\infty}\hat{g}_n = \lim_{n\to\infty}\frac{\hat{g}_n\hat{\gamma}_m}{\hat{\gamma}_m} = \lim_{n\to\infty}\frac{\hat{g}_m\hat{\gamma}_n}{\hat{\gamma}_m}$$

$$=\frac{\hat{g}_m}{\hat{\gamma}_m}\lim_{n\to\infty}\hat{\gamma}_n = \frac{\hat{g}_m}{\hat{\gamma}_m} \text{ on } U. \tag{178}$$

Since $\hat{\gamma}_m\in S$ and $\hat{\gamma}_m>0$ on U. $\hat{G}$ is an infinitely differentiable function on U, this completes the proof of (ii).

(iii) Let $\varphi\in\mathcal{D}$. Then there exists $m\in N$ such that $\hat{\gamma}_m>0$ on the support of φ. We, then, write

$$(F*G)^\wedge(\varphi) = \lim_{n\to\infty}(f_n*g_n)^\wedge(\varphi)$$

$$=\lim_{n\to\infty}\hat{f}_n(\hat{g}_n\varphi)$$

$$=\hat{G}\lim_{n\to\infty}(\hat{f}_n\hat{\gamma}_n)(\varphi)$$

$$= \hat{G} \lim_{n\to\infty} (f_n * \hat{\gamma}_n)^\wedge (\varphi)$$

$$= (\hat{F}\hat{G})(\varphi)$$

i.e. $$(F * G)^\wedge (\varphi) = (\hat{F}\hat{G})(\varphi) \tag{179}$$

The proof is completed on using (i).

(iv) Let $\varphi \in \mathcal{D}, \forall m \in N$

$$(\hat{F}\hat{\varphi}_m)(\varphi) = \hat{F}(\hat{\varphi}_m \varphi)$$

$$= \lim_{n\to\infty} \hat{f}_n (\hat{\varphi}_n \varphi)$$

$$= \lim_{n\to\infty} (\hat{f}_n \hat{\varphi}_m)(\varphi)$$

i.e. $$(\hat{F}\hat{\varphi}_m)(\varphi) = \hat{f}_m(\varphi) = \hat{f}_m \tag{180}$$

This completely proves the theorem.

5.3 Cauchy Representation of Hilbert Transform for Integrable Boehmians [26]

Relation of Cauchy representation with the Hilbert transform on L_2 spaces are described in this section. Further, by virtue of the analytic representation of Hilbert transform on the space of Schwartz distribution, we have investigated it for integrable Boehmians.

5.3.1 Prerequisites : Hilbert transform, known for its application in the theory of generalized functions, of a function *f* is defined as the Cauchy principal value of the integral [cf. Orton [92]] by

$$H\{f(t)\} = -\frac{1}{\pi} \int_{-\infty}^{\infty} \frac{f(t)}{(t-x)} dt. \tag{181}$$

We notice, also, that the distributional Hilbert transformation is defined in terms of analytic representations of distributions, which replaces the Cauchy integrals used in the classical theory.

Definition 2 : Let $F \in \mathcal{D}'$. Then the analytic representation of *f* is defined to be any function $\hat{f}(z)$, which is analytic on the complement of the support of *f* such that for all test functions $\varphi \in \mathcal{D}$,

$$\lim_{y\to 0}\int_{-\infty}^{\infty}[\hat{f}(x+iy)-\hat{f}(x-iy)]\varphi(x)dx=\langle f(x),\varphi(x)\rangle. \tag{182}$$

Theorem 10 : [92, p.657]: Let $F \in \mathcal{D}'$ is a distribution of compact support. Then the analytic representation of f is given by the generalized Cauchy integral

$$\hat{f}(z)=\frac{1}{2\pi i}\left\langle f(t),\frac{1}{t-z}\right\rangle, \tag{183}$$

where $z = x + iy$ and $y \neq 0$. For a given $f, \hat{f}(z)$ is determined uniquely within the space of functions, which are analytic for $y \neq 0$ and satisfy

$$\left|\hat{f}(z)\right| \leq M|y|^{-n}, 0<|y|<1 \tag{184}$$

For some integer *n* and a constant *M*.

Definition 3: Let $F \in \mathcal{D}'$ and $\hat{f}(z)$ is an analytic representation of f . Then the Hilbert transform of *f*, relative to $\hat{f}(z)$, is defined as the distributional limit

$$H\hat{f}(x)=\lim_{y\to 0} i[\hat{f}(x+iy)+\hat{f}(x-iy)] \tag{185}$$

$$=\frac{1}{\pi}f(x)*\text{p.v.}\frac{1}{x}. \tag{186}$$

Definition 4 : The space $\mathcal{D}'_+$ is the space of all distributions $f \in \mathcal{D}'$ (R) such that $f(x)=\hat{f}_+(x)$ for some analytic (Cauchy) representation $\hat{f}(z)$ of *f*. The space $\mathcal{D}'_-$ is the space of all distributions $f \in \mathcal{D}'$ (R) for which $f(x)=-\hat{f}_-(x)$ for some analytic representation $\hat{f}(z)$.

Now we proceed to investigate integrable Boehmians for the Cauchy representation of Hilbert transform.

5.3.2 Main Results

Lemma 2 : If $[f_n/\delta_n] \in B_{L_1}$, then the sequence

$$H\hat{f}_n(x)=\frac{1}{\pi}\int_{-\infty}^{\infty}\frac{f(x)}{(y-x)}dx$$

converges uniformly on each compact set in R.

Proof : If (δ_n) is a delta sequence, then $[H(\hat{\delta}_k)]$ converges on each compact set to the function 1. Thus, for each compact set K, $[H(\hat{\delta}_k)] > 0$ on K for almost all $k \in K$, and

$$H\hat{f}_n(x) = \frac{(H\hat{f}_n)(H\hat{\delta}_k)}{(H\hat{\delta}_k)} = \frac{H*(\hat{f}_n * \hat{\delta}_k)}{H(\hat{\delta}_k)}$$

$$= \frac{H(\hat{f}_k)H(\hat{\delta}_n)}{H(\hat{\delta}_k)} \text{ on } K.$$

The conclusion of this is same, that is why we avoid, as Lemma 1 of Section 5.2.2.

Theorem 11 : Let $F, G \in B_{L_1}$ .Then

(i) $H(\lambda F) = \lambda H(F)$, for any complex number λ, and

$H(F+G) = H(F) + H(G)$

(ii) $H(F * G) = H(F)H(G)$

(iii) If $\Delta - \lim_{n\to\infty} F_n = F$, then $H(F_n) \to H(F)$

uniformly on each compact set.

Proof : From Definitions 2 and 3, we notice that the proofs of (i) and (ii) readily follow. We prove (iii).

We have

$$\delta - \lim_{n\to\infty} F_n - F \Rightarrow H(F_n) \to H(F)$$

uniformly on each compact set. Let (δ_n) is a delta sequence such that

$$F_n * \delta_k, F * \delta_k \in L_1, \forall n, k \in N$$

and

$$\|(F_n - F) * \delta_k\| \to 0 \text{ as } n \to \infty \text{ for each } k \in N.$$

Let K be a compact set in R. Then $H(\delta_k) > 0$ on K for some $k \in N$. Since $H(\delta_k) > 0$ is continuous, we have

$$H(F_n).H(\delta_k) \to H(F).H(\delta_k)$$

uniformly on K. Finally, for

$$H(F_n).H(\delta_k) - H(F).\delta_k = H((F_n - F) * \delta_k)$$

and the above mentioned norm, the theorem is proved.

On Inversion :

We notice [92,p.663] that the analytic representation of f is given by equation (183) and the Hilbert transform of it is given by (186). Applying the convolution theorem for the distributional Fourier transform, we get

$$F\left(H\hat{f}\right)(s) = F\left(-\frac{1}{\pi}\text{p.v.}\frac{1}{x}\right)(s)F(f(s))$$

i.e.
$$F\left(H\hat{f}\right)(s) = -i\,\text{sgn}(s)F(f(s)). \tag{187}$$

This, indeed, justifies that the Fourier transform of the Cauchy representation of the Hilbert transform is a Fourier transform. Then by virtue of the properties of the Fourier transform and the relation between the Fourier transform and the Hilbert transform for the Cauchy representation, the inversion formula for integrable Boehmian and tempered Boehmian can be established, see Loonker and Banerji [55]

5.4 Fourier-Bessel Transform for Tempered Boehmians [24]

In this section we have introduced the tempered Boehmian as an extension of the tempered distribution. We have studied its properties by invoking relation between Fourier and Hankel transform.

5.4.1 Definitions: Hankel transform, at times, used as synonym with Fourier-Bessel transform. The Fourier transform of a function with respect to each of the n-independent variables is defined, see Sneddon [103], by

$$F_{(n)}(\xi_1, \xi_2, \ldots\ldots, \xi_n)$$
$$= F_n[f(x_1, x_2, \ldots\ldots, x_n); x_1 \to \xi_1, \ldots.., x_n \to \xi_n]. \tag{188}$$

Let us consider X and ξ by n-vectors $(x_1, x_2, \ldots\ldots, x_n)$ and $(\xi_1, \xi_2, \ldots\ldots, \xi_n)$ that has the inner product $x_1\xi_1 + x_2\xi_2 + ..x_n\xi_n = X.\xi$. Then the n-dimensional Fourier transform

$$F_{(n)}(\xi) = F_{(n)}[f(X); x \to \xi] \tag{189}$$

of $f(X)$ is defined by

$$F_{(n)}(\xi) = (2\pi)^{-\frac{n}{2}} \int_{E_n} f(x) e^{i(\xi.X)} dX, \tag{190}$$

where E_n is the n-dimensional Euclidean space. The inversion formula for the Fourier transform of $f(X)$ and its convolution of $f(X)$ and $g(X)$ of the vector X are defined, respectively, by

$$f(X) = (2\pi)^{-\frac{n}{2}} \int_{E_n} f_{(n)}(\xi) e^{-i(\xi.X)} d\xi \tag{191}$$

and

$$F_{(n)}[fog(x); X \to \xi] = F_{(n)}(\xi) G_{(n)}(\xi). \tag{192}$$

For two dimensional Fourier transform in plane polar coordinates (r, θ) in x_1x_2 − plane and (ρ, φ) in $\xi_1\xi_2$ − plane, we have $x_1 = r\cos\theta$, $x_2 = r\sin\theta, \xi_1 = \rho\cos\varphi$, and $\xi_2 = \rho\sin\varphi$. This yields $(X.\xi) = r\rho\cos\ (\theta - \varphi)$, and then we obtain

$$F_{(2)}[f(r), x_1 \to \xi_1, x_2 \to \xi_2]$$

$$= \frac{1}{2\pi} \int_0^\infty r dr \int_0^{2\pi} f(r) e^{ir\rho\cos(\theta-\varphi)} d\theta. \tag{193}$$

From Watson [112], we have

$$J_0(r\rho) = \frac{1}{2\pi} \int_0^{2\pi} e^{ir\rho\cos(\theta-\varphi)} d\theta \tag{194}$$

such that (193) is expressed as

$$F_{(2)}[f(r), x_1 \to \xi_1, x_2 \to \xi_2] = \int_0 r f(r) J_0(r\rho) dr. \tag{195}$$

In other words,

$$F_{(2)}[f(r), x_1 \to \xi_1, x_2 \to \xi_2] = H_0[f(r); \rho], \tag{196}$$

where H_0 is the Hankel transform of order zero.

We observe the evidence [see[100]] of viewing the Hankel transform of order zero as two dimensional Fourier transform of circularly symmetric functions. Since the Fourier transform for the tempered distribution and tempered Boehmian is established in [78] and [77], we have used the relation between two dimensional Fourier transform and the Hankel transform to establish a result for Boehmians of slow growth.

An Illustration: The distribution space of slow growth S', for the Fourier transform, is defined in terms of the Parseval equation by, see [38]

$$\left\langle \tilde{f}, \tilde{\varphi} \right\rangle = 2\pi \langle f, \varphi \rangle \tag{197}$$

and

$$\left\langle f, \bar{\tilde{\varphi}} \right\rangle = 2\pi \langle f, \tilde{\varphi} \rangle,$$

respectively. Similarly, using relation between two-dimensional Fourier transfom and the Hankel transform, the tempered distribution space S' for the Hankel transform is given by

$$\langle \hat{g}(\rho), \hat{\varphi}(\rho) \rangle = \langle g(r), \varphi(r) \rangle \tag{198}$$

In other words,

$$\langle F_2(f), F_2(\varphi) \rangle = \langle f, \varphi \rangle$$

i.e.

$$\langle H_0(f), H_0(\varphi) \rangle = \langle f, \varphi \rangle. \tag{199}$$

5.4.2 The Main Result :

Theorem 12: If $[f_n / \varphi_n] \in B_l$, then the sequence $H_0\{f_n\}, n = 1, 2, \ldots, \infty$ converges in $\mathcal{D}'$. Moreover, if $[f_n / \varphi_n] = [g_n / \psi_n]$, then $H_0\{f_n\}$ and

$H_0\{g_n\}$ converge to the same limit for the Hankel transformation of tempered Boehmians.

Proof : Let $\varphi \in \mathcal{D}$ (the testing function space) and $k \in N$ be such that $H_0\varphi_k > 0$ on the support of φ. Then

$$\langle H_0\{f_n\}, \varphi_n \rangle = \left\langle H_0\{f_n\}, \varphi . \frac{H_0(\varphi_k)}{H_0(\varphi_k)} \right\rangle$$

$$= \left\langle H_0\{f_n\} . H_0\{\varphi_k\}, \frac{\varphi}{H_0\{\varphi_k\}} \right\rangle$$

$$= \left\langle H_0\{f_k\} . H_0\{\varphi_n\}, \frac{\varphi}{H_0\{\varphi_k\}} \right\rangle$$

i.e.

$$= \left\langle H_0\{f_k\}, \frac{\varphi . H_0\{\varphi_n\}}{H_0\{\varphi_k\}} \right\rangle$$

Since the sequence $\left\{ \frac{\varphi, H_0\{\varphi_n\}}{H_0\{\varphi_n\}} \right\}$ converges to $\frac{\varphi}{H_0\{\varphi_k\}}$ in $\mathcal{D}$, the sequence $\{H_0\{f_n\}, \varphi\}$ converges. This proves that the sequence $H_0\{f_n\}$ converges in $\mathcal{D}'$ (dual of the space $\mathcal{D}$).

Now, assume that $[f_n / \varphi_n] = [g_n / \psi_n] \in B_I$. Let we define those that we have mentioned in the proof of Theorem 8. This will imply $H_0\{h_n\}$ converges in $\mathcal{D}'$. Moreover, on the same line of analysis the theorem will be proved.

Definition 5 : Let $F = [f_n / \varphi_n] \in B_I$ and the limit of sequence $H_0\{f_n\}$ is in $\mathcal{D}'$. The Hankel transform $H_0\{F\}$ of F, thus, has a mapping from B_I into $\mathcal{D}'$, which is linear.

Chapter 6

Hartley Transformation for Boehmians

6.1 Introduction

An integral transformation, which maps a real-valued temporal of special functions into a real-valued frequency function through the kernel $cas(vx) = \cos(vx) + \sin(vx)$, is called the Hartley transform. This symmetrical formulation of the traditional Fourier transform, which is attributed to Ralph Vinton Lyon Hartley [40], leads to a parallelism that exists between the function of the original variable and that of its transform. Hartley transform allows a function to be decomposed into two independent sets of sinusoidal components, which are represented in terms of positive and negative frequency components, repectively.

This is in contrast to the complex exponential, exp (*jwx*), used in classical Fourier analysis. Perhaps one of Hartley's most outstanding contribution (the transform under study) was more a symmetrical Fourier integral, which was originally developed for the steady-state and transient analysis of telephone transmission problem. After forty years of quiescence, it was rediscovered by Wang [109,110,111] and Bracewell [10,11,12]. Hartley transform of a function *f*(*x*) can be expressed either as

$$H(v) = \frac{1}{\sqrt{2\pi}} \int_{-\infty}^{\infty} f(x)cas(vx)dx, \tag{200}$$

or as

$$H(f) = \int_{-\infty}^{\infty} f(x) cas(2\pi f x) dx, \tag{201}$$

where the angular or radian frequency variable v is related to the frequency variable f by $v = 2\pi f$ and

$$H(f) = \sqrt{2\pi}\ H(2\pi f) = \sqrt{2\pi}\ H(v). \tag{202}$$

The integral kernel, called the *cosine* and *sine* functions or *cas* function, is defined by

$$cas(vx) = \cos(vx) + \sin(vx).$$

$$= \sqrt{2}\left\{\cos\frac{\pi}{4}\cos vx + \sin\frac{\pi}{4}\sin vx\right\}$$

$$= \sqrt{2}\cos\left(vx - \frac{\pi}{4}\right) \tag{203}$$

or in other form,

$$\cos(vx) = \sqrt{2}\sin\left(vx + \frac{\pi}{4}\right). \tag{204}$$

The inverse Hartley transform may be defined by either

$$f(x) = \frac{1}{\sqrt{2\pi}}\int_{-\infty}^{\infty} H(v) cas(vx) dx, \tag{205}$$

or

$$f(x) = \int_{-\infty}^{\infty} H(f) cas(2\pi f x) df. \tag{206}$$

The angular velocity v, with units of radian per second, is equivalent to the frequency variable w in the Fourier domain. However, it is used here to distinguish $H(v)$, the Hartley transform of $f(x)$, from the Fourier transform $f(w)$ of $f(x)$. From (200) and (201), it is noticed that the inverse transformation (synthesis equation) requires the identical integral equation as the direct transformation. Hartley transform has the convenient property of being its own inverse (the involution), that is

$$f = \{H(Hf)\}. \tag{207}$$

Details, with regard to Hartley transform, can be seen in Olejniczak [19] and [39]

6.2. Hartley Transform for Integrable Boehmians [56]

Basic terminologies required in this section, wherever needed, will be referred to Section 6.1. Relation between the Hartley transform and the Fourier transform is given by

$$H(v) = [R[F(\omega)] - I[F(\omega)]]_{\omega=v}, \tag{208}$$

where R and I denote, respectively, the real and imaginary parts of the Fourier transform. Alternatively, the Fourier transform is expressed in terms of the Hartley transform as

$$F(\omega) = [H^e(v) - jH^\circ(v)]_{v=\omega}, \tag{209}$$

where $H^e(v)$ and $H^\circ(v)$ are, respectively, the even and odd parts of the Hartley transform.

Since the Hartley transform is a symmetrical representation of the Fourier transform, the convergence conditions for the same (the former) will be satisfied for the L_p – space and the function $f(x)$, which is absolutely integrable for $\int_{-\infty}^{\infty} |f(x)| dx < \infty$. The function $f(x)$ has a finite number of discontinuities over every finite interval as well as finite number of local maximum and local minimum points over any finite interval. Properties of the Hartley transform [15] are

(1) **Linearity :** If $f_1(x)$ and $f_2(x)$ possess the Hartley transforms $H_1(f)$ and $H_2(f)$, respectively, then

$$\int_{-\infty}^{\infty} [\alpha f_1(x) + \beta f_2(x)] cas(2\pi fx) = \alpha H_1(f) + \beta H_2(f). \tag{210}$$

(2) **Scaling :** If $f(x)$is the Hartley transform, given in (201), then

$$\int_{-\infty}^{\infty} f(kx)cas(2\pi fx)dx = \frac{1}{k}H\left(\frac{f}{k}\right). \tag{211}$$

(3) **Convolution :** If $f_1(x)$ and $f_2(x)$ have Hartley transforms $H_1(f)$ and $H_2(f)$, respectively, then

$$f_1(x) * f_2(x) = \frac{1}{2}[H_1(f)H_2(f) + H_1(-f)H_2(f) + H_1(f)H_2(-f) - H_1(-f)H_2(-f)] \tag{212}$$

for its symmetry following conditions hold.

(i) If $f_1(x)$ and (or) $f_2(x)$ (resp. are (is)) even, or $f_1(x)$ is even and $f_2(x)$ is odd or vice-versa, then

$$f_1(x) * f_2(x) = H_1(f)H_2(f). \tag{213}$$

(ii) If $f_1(x)$ is odd, then

$$f_1(x) * f_2(x) = H_1(f)H_2(-f), \tag{214}$$

and if $f_2(x)$ is odd, then

$$f_1(x) * f_2(x) = H_1(-f)H_2(f). \tag{215}$$

whereas, if both functions are odd, then

$$f_1(x) * f_2(x) = -H_1(f)H_2(f). \tag{216}$$

(4) **Derivative :** The *n*th derivative of $f(x)$ is

$$f^{(n)}(x) = cas'\left(\frac{n\pi}{2}\right)(2\pi f)^n H[(-1)^n f]. \tag{217}$$

(5) **Parseval relation :** If for $f_1(x)$ and $f_2(x)$, Hartley transforms are $H_1(f)$ and $H_2(f)$, respectively, then

$$\int_{-\infty}^{\infty} f_1(x)f_2(x)dx = \int_{-\infty}^{\infty} H_1(f)H_2(f)df. \tag{218}$$

6.2.1 Main Results :

The results, investigated in this section, presumably, appeared for the first time.

Lemma 1 : If $[f_n/\delta_n] \in B_{L_1}$, then the sequence

$$H(v) = H[f_n](v) = \frac{1}{\sqrt{2\pi}} \int_{-\infty}^{\infty} f_n(x) cas(vx) dx$$

or

$$H(f_n) = \int_{-\infty}^{\infty} f_n(x) cas(2\pi f x) dx$$

converges uniformly on each compact set in *R*.

Proof : If (δ_n) is a delta sequence, then $H(\delta_n)$ converges uniformly on each compact set to a constant function 1. Hence, for each compact set $K, H(\delta_k) > 0$ on *K* and for almost all $k \in K$. Further,

$$H(f_n) = H(f_n)\frac{H(\delta_k)}{H(\delta_k)} = \frac{H(f_n * \delta_k)}{H(\delta_k)},$$

that is,

$$= \frac{H(f_k * \delta_n)}{H(\delta_k)} = \frac{H(f_k)}{H(\delta_k)} H(\delta_n) \text{ on } K.$$

Hence the proof is completed.

Theorem 1 : Let $F, G \in B_{L_1}$. Then following hold:

(i) $H(\alpha F + \beta G) = \alpha H(F) + \beta H(G)$

(ii) $H(F(kx)) = (1/k) H(f/k)$

(iii) $H(F * G) = H(F).H(G)$

(iv) $H[F^{(n)}(x)] = cas'\left(\frac{n\pi}{2}\right)(2\pi f)^n H[(-1)^n f]$

(v) If $H(f) = 0$, then $F = 0$

(vi) If $\Delta - \lim_{n\to\infty} f_n = F$, then $H(F_n) \to H(F)$ uniformly on each compact set.

Proof : By invoking (210)-(218), we can prove (i)-(iv). The analysis being simple, the details are avoided.

To prove (v), we consider the uniqueness theorem for the Hartley transform for the generalized function [15], which indeed shows that *f-g* is a zero distribution, and *f* = *g*, whence

$$\langle f-g,\varphi\rangle=\langle H(f-g),H(\varphi)\rangle$$
$$=\langle H(f)-H(g),H(\varphi)\rangle=\langle F-G,H(\varphi)\rangle=0. \qquad (219)$$

This completes the proof of (v).

Now to prove (vi), we know that

$$\delta-\lim_{n\to\infty} F_n=F \Rightarrow H(F_n)\to H(F)$$

uniformly on each compact set. Let (δ_n) has the usual meaning such that

$$F_n*\delta_k, F*\delta_k \in L_1, \quad \text{for all } n,k\in N,$$

and $\|(F_n-F)*\delta_k\|\to 0$ as $n\to\infty$, for $n,k\in N$, and K is a compact set in R. Then $H(\delta_k)\to 0$ on K for some $k\in N$,

Since $H(\delta_k)$ is a continuous function, we have

$$H(F_n).H(\delta_k)\to H(F)H(\delta_k)$$

uniformly on K. Further, since

$$H(F_n).H(\delta_k)-H(F).H(\delta_k)=H((F_n-F)*\delta_k)$$

and $\|(F_n-F)*\delta_k\|\to 0$ as $n\to\infty$, which proves (vi) and the theorem is completely proved.

Lemma 2 : If $f\in L_1$ and

$$f_n(x)=\frac{1}{\sqrt{2\pi}}\int_{-\infty}^{\infty} H(v)cas(vx)dv$$

or

$$(f_n)=\int_{-\infty}^{\infty} H(f)cas(2\pi fx)df,$$

then (f_n) converges to f in L_1 – norm.

Theorem 2 : If the expression in Lemma 2 holds, then

$\delta-\lim_{n\to\infty} f_n=F$ (hence, also $\Delta-\lim_{n\to\infty} f_n=F$).

Proof : Let $F=[g_n/\varphi_n]$ and $k\in N$. Then employing relation (218), we obtain

$$\int_{-\infty}^{\infty} H(f_n)H(\delta_k)df=\int_{-\infty}^{\infty} f_n(x)\delta_k(x)dx$$

$$= \int_{-\infty}^{\infty} H(f_n) \int_{-\infty}^{\infty} \delta_k(x) cas(2\pi fx) df dx$$

$$= \int_{-\infty}^{\infty} \delta_k(x) dx \int_{-\infty}^{\infty} H(f_n) cas(2\pi fx) df$$

i.e. $$= \int_{-\infty}^{\infty} f_n(x)\delta_k(x) dx.$$

By Lemma 2 and using the definition of convergence, $\|(f_n * \delta_k) - f * \delta_k\| \to 0$ as $n \to \infty$. Since k is an arbitrary positive integer, $\delta - \lim_{n\to\infty} f_n = F$. This theorem, similarly, can be proved for $\Delta - \lim_{n\to\infty} f_n = F$.

6.3 Hartley Transform for Tempered Boehmians[52]

Theorem 3 : If $[f_n / \varphi_n] = B_l$, then the sequence $\{Hf_n\}$ converges in $\mathcal{D}'$. Moreover, if $[f_n / \varphi_n] = [g_n / \gamma_n] \in B_l$, then the sequence $\{Hf_n\}$ and $\{Hg_n\}$ converge to same limit.

Proof : Owing to much similarity with the Theorem 8, Section 5.2.3 and the Theorem 12, Section 5.4.2, we take liberty of skipping detailed proof.

Let $\varphi \in \mathcal{D}$ and $k \in N$. We have $Hf_n . H\varphi_m = Hf_m . H\varphi_n$.

Hence

$$\langle H(f_n), \varphi \rangle = \left\langle Hf_n, \frac{\varphi H\varphi_k}{H\varphi_k} \right\rangle$$

i.e. $$= \left\langle Hf_k, \frac{\varphi H\varphi_n}{H\varphi_k} \right\rangle.$$

Let $[f_n / \varphi_n] = [g_n / \gamma_n] \in B_l$ and h_n and δ_n have same meaning as that of Theorem 8, Section 5.2.3. we conclude then that $[h_n / \delta_n] = [f_n / \varphi_n]$ $= [g_n / \gamma_n]$ and the sequence $\{Hh_n\}$ converges in $\mathcal{D}'$. Moreover,

$$\lim_{n\to\infty} \langle Hh_{2n-1}, \varphi \rangle = \lim_{n\to\infty} \langle Hh_n, \varphi \rangle$$

proves that the sequences $\{Hf_n\}$ and $\{Hh_n\}$ have same limit.

Theorem 4: Let $F=[f_n/\varphi_n]=B_l$ and $G=[g_n/\gamma_n]\in B_S$

Then

(i) $f^{(k)}(x)=cas'\dfrac{n\pi}{2}(2\pi f)^k H[(-1)^k f]$, where

$$k=1,\frac{\partial f}{\partial x}=(2\pi f)H[-f]$$

(ii) HG has infinitely differentiable function

(iii) $H(F*G)=H(F).H(G)$.

Proof : We have

$$\left(\frac{\partial f}{\partial x}\right)=H\left[\left(\frac{\partial f}{\partial x}*\varphi_n\right)/(\varphi_n*\varphi_n)\right]$$

$$=\lim_{n\to\infty}H\left(\frac{\partial f}{\partial x}*\varphi_n\right)$$

$$=\lim_{n\to\infty}(2\pi f_n)H[-f_n]$$

i.e. $$=(2\pi f)[H[-f]].$$

Here the last equality follows from Theorem 3 and from $(f_n/\varphi_n)=[(f_n*\varphi_n)/(\varphi_n*\varphi_n)]$.

(ii) [We may refer the analysis of this part to Theorem 5, Section 5.1. But we do not skip, here, any step of analysis]

Let $[g_n/\gamma_n]\in B_S$ and U be the bounded open subset of R^N. Then there exists $m\in N$ such that $\{H\gamma_m\}>0$ on U and

$$H(G)=\lim_{n\to\infty}\{Hg_n\}=\lim_{n\to\infty}\frac{\{Hg_n\}\{H\gamma_m\}}{\{H\gamma_m\}}$$

$$=\frac{\{Hg_m\}}{\{H\gamma_m\}}\lim_{n\to\infty}\{H\gamma_n\}$$

i.e. $$=\frac{\{Hg_m\}}{\{H\gamma_m\}}\text{ on }U.$$

Since $\{Hg_m\},\{H\gamma_m\}\in S$ and $\{H\gamma_m\}>0$ on U, $\{HG\}$ is proved to be an infinitely differentiable function on U.

(iii) Let $F=[f_n/\varphi_n]=\in B_I$ and $G=[g_n/\gamma_n]=\in B_S$. If $\varphi\in \mathcal{D}$, then $\{H\gamma_n\}>0$ on the support of φ, and

$$H(F*G)(\varphi)=\lim_{n\to\infty}H(f_n*g_n)(\varphi)$$

$$=\lim_{n\to\infty}H\{(Hf_n.Hg_n)\}(\varphi)=\lim_{n\to\infty}(Hf_n)(Hg_n,\varphi)$$

i.e. $$=(HG)\lim_{n\to\infty}(Hf_n)(\varphi(H\gamma_n))$$

i.e. $$=(HF.HG)(\varphi).$$

The theorem is, thus, proved.

Lemma 3: If $F=[f_n/\varphi_n]=\in B_I$, then $HF.H\varphi_m=Hf_m$, for all $m\in N$.

Proof : Let $\varphi\in \mathcal{D}$. Then

$$\langle HF.H\varphi_m,\varphi\rangle=\langle HF,H\varphi_m.\varphi\rangle$$

$$=\lim_{n\to\infty}\langle Hf_n,H\varphi_m.\varphi\rangle$$

$$=\lim_{n\to\infty}\langle Hf_n.H\varphi_m,\varphi\rangle$$

$$=\lim_{n\to\infty}\langle Hf_m.H\varphi_n,\varphi\rangle$$

i.e. $$\langle HF.H\varphi_m,\varphi\rangle=\langle Hf_m,\varphi\rangle.$$

Hence, $$HF.H\varphi_m=Hf_n.$$

This completes the proof of the lemma.

Theorem 5 : The distribution f is the Hartley transform of a tempered Boehmian if and only if there exists a delta sequence $\{\delta_n\}$ such that it is a tempered distribution for every $n\in N$.

Proof : From Lemma 3, if $F=[f_n/\varphi_n]=\in B_I$, then $f.H\varphi_n=HF.H\varphi_n$ $=Hf_n$. Hence, $f.H\varphi_n$ is a tempered distribution. Conversely, if $f\in \mathcal{D}'$ and $\{\delta_n\}$ is a delta sequence, then $\{fH\delta_n\}$ is a tempered distribution for every $n\in N$. Thus, we have

$$F=\left[\frac{(H^{-1}(fH\delta_n)*\delta_n)}{(\delta_n*\delta_n)}\right], \tag{220}$$

where $H^{-1}(f\ H\delta_n)$ is the inverse Hartley transform of $(f\ H\delta_n)$. Since $HF = f$, therefore F is proved to be a tempered Boehmian.

Theorem 6: (Inversion): Let F be a tempered Boehmain and $HF = f$. Then F is, the one given by the relation (220).

Proof: If $F = [f_n / \varphi_n]$ and $\delta_n = \varphi_n$ are considered, then the required inversion formula is given by

$$F = \left[\frac{H^{-1}(fH\varphi_n)}{(\varphi_n)}\right].$$

Details are avoided because the proof is similar to that of Theorem 5.

6.4 Tempered Distributional Hartley Transform [52]

With regard to Zemanian [115 and 116], spaces S and S', which are employed in the investigations in this section, can be referred to for definitions in the preceding sections of this monograph. We have used Bracewell [11] and Choudhary and Thorat [15]. Applicaton of the Hartley transform on testing function space of descent is investigated in this section, which is followed by the use of the Hartley transform on the corresponding dual space (or the conjugate space). Terminologies used can be referred to in this chapter in the preceding section.

6.4.1 Main Results

Theorem 7 : The Hartley transform $\hat{f}(s)$ and its inverse $f(t)$ map the space S into itself.

Proof : In slightly different notations, than those in (200), we write

$$\hat{f}(s) = h[f](s) = \int_{-\infty}^{\infty} f(t) cas(2\pi st) dt \tag{221}$$

to define the Hartley transform. The inverse of it will be given by

$$f(t) = h^{-1}[f](t) = \int_{-\infty}^{\infty} \hat{f}(s) cas(2\pi st) ds, \tag{222}$$

where s is real and the kernel is

$$cas(2\pi st) = \cos(2\pi st) + \sin(2\pi st), \tag{223}$$

[which are same as given in (203) – (206)] such that

$$cas(2\pi st) = \left(\frac{1-i}{2}\right)e^{i2\pi st} + \left(\frac{1+i}{2}\right)e^{-i2\pi st}. \tag{224}$$

Now, let $\varphi(t) \in S$. Then its Hartley transform, given by (as a consequence of relation (224))

$$\hat{f}(s) = \int_{-\infty}^{\infty} f(t)\left[\left(\frac{1-i}{2}\right)e^{i2\pi st} + \left(\frac{1+i}{2}\right)e^{-i2\pi st}\right]dt, \tag{225}$$

is differentiated k-times with respect to s, we get

$$\hat{\varphi}^{(k)}(s) = (2\pi i)^k \int_{-\infty}^{\infty} [t^k \varphi(t)]$$

$$\times\left[\left(\frac{1-i}{2}\right)e^{i2\pi st} + (-1)^k\left(\frac{1+i}{2}\right)e^{-i2\pi st}\right]dt. \tag{226}$$

For the sake of simplicity, we let

$$C_1 = \frac{1-i}{2} \text{ and } C_2 = (-1)^k\left(\frac{1+i}{2}\right)$$

both being constants. When, with C_1 and C_2 written, the relation (226) is integrated by parts m-times, we get

$$\hat{\varphi}^{(k)}(s) = \frac{(2\pi i)^k(-1)^m}{(2\pi is)^m}\int_{-\infty}^{\infty}\frac{d^m}{dt^m}[t^k\varphi(t)]\left[C_1e^{i2\pi st} + C_2e^{-i2\pi st}\right]dt,$$

i.e. $$\left|s^m\hat{\varphi}^{(k)}(s)\right| \le (2\pi)^{k-m}\int_{-\infty}^{\infty}\left|\frac{d^m}{dt^m}[t^k\varphi(t)]\right|\left|C_1e^{i2\pi st} + C_2e^{-i2\pi st}\right|dt. \tag{226}$$

Where, in (226), we have

$$\left|C_1e^{i2\pi st} + C_2e^{-i2\pi st}\right|$$

$$= \left|\left(\frac{1-i}{2}\right)e^{i2\pi st} + (-1)^k\left(\frac{1+i}{2}\right)e^{-i2\pi st}\right|$$

$$\le \left|\frac{1-i}{2}\right| + \left|\frac{1+i}{2}\right| = \sqrt{2}. \tag{228}$$

Using the relation (228) in (226), we get

$$\left|s^m P\hat{\varphi}^{(k)}(s)\right| \le (2\pi)^{k-1}\sqrt{2}\int_{-\infty}^{\infty}\left|\frac{d^m}{dt^m}[t^k\varphi(t)]\right| dt. \tag{229}$$

We invoke, now, the fact that if $\varphi(t)\in S$, then $t^k(\varphi)(t)\in S$ and the integral in right–hand of (229) is bounded (say, by a constant C_{mk}, which depends on non- negative integers *m* and *k*). Hence,

$$\left|s^m\hat{\varphi}^{(k)}(s)\right| \le (2\pi)^{k-m}\sqrt{2}C_{mk} = G_{mk}$$

i.e.

$$\left|s^m\hat{\varphi}^{(k)}(s)\right| \le G_{mk}, -\infty < S < \infty \tag{230}$$

for every pair of non-negative integers *m* and *k*. Thus, we have shown that if $\varphi(t)\in S$, then its Hartley transform $\hat{\varphi}(s)$ is also in *S*. Similarly, we can prove that the inverse Hartley transform also maps the space into itself.

Theorem 8 : The Hartley transform and its inverse are continuous linear mapping of *S* into itself.

Proof : In Section 6.2, we have mentioned the property of linearity for the Hartley transform and its inverse. What now, in this theorem, remains is to show its continuity. We write following.

Let the sequence $\{\varphi_p\}_{p=1}^{\infty}$ converges in δ to zero. We intend to show that $\{\hat{\varphi}_p\}_{p=1}^{\infty}$ also converges in *S* to zero. Invoking Leibniz theorem in the right-hand side of (229), we get

$$\left|s^m\hat{\varphi}_p^k(s)\right| \le (2\pi)^{k-m}\sqrt{2}\int_{-\infty}^{\infty}\left|\sum_{r=0}^{m}\binom{m}{r}\frac{dr}{dt^r}(t^k)\frac{d^{m-r}}{dt^{m-r}}(t)[\varphi_p(t)]\right| dt$$

i.e. $$\le (2\pi)^{k-m} r!\sqrt{2}\sum_{r=0}^{\min(m,k)}\binom{m}{r}\binom{k}{r}\int_{-\infty}^{\infty}\frac{\left|(1+t^2)t^{k-r}\varphi_p^{(m-r)}(t)\right|}{\left|1+t^2\right|}dt. \tag{231}$$

Now, by virtue of

$$\left|t^m\varphi_p^{(k)}(t)\right| \le C_{mk}, -\infty < t < \infty$$

we notice $\left|(1+t^2)t^{k-r}\varphi_p^{(m-r)}(t)\right|$ to be bounded, say, by a constant C_{mkrp}. Hence, (231) implies

$$\left|s^m \hat{\varphi}_p^{(k)}(s)\right| \le \pi(2\pi)^{k-m} r!$$

$$\times\sqrt{2}\sum_{r=0}^{\min(m,k)}\binom{m}{r}\binom{k}{r}C_{mkrp}. \qquad (232)$$

As $P \to \infty, \left|s^m \hat{\varphi}_p^{(k)}(s)\right| \to 0$, that is, $\{\hat{\varphi}_p\}_{p=1}^{\infty}$ converges in δ to zero.

6.5 Hartley Transform of Distributions of Slow Growth [82]

Parseval's equation (see equation (218)) provides the definition for the Hartley transform of distributions of slow growth. If the locally integrable function *f*(*t*) is absolutely integrable for $-\infty < t < \infty$ and φ is a testing function for rapid descent, then its Hartley transform exists and produces a Parseval's relation

$$\langle f, \hat{\varphi}\rangle = \int_{-\infty}^{\infty} f(s)\hat{\varphi}(s)ds$$

i.e.

$$= \int_{-\infty}^{\infty} ds \int_{-\infty}^{\infty} f(s)\varphi(t)cas(2\pi st)dt. \qquad (233)$$

Changing the order of integration, due to the absolute convergence of integrals, we obtain

$$\langle f, \hat{\varphi}\rangle = \langle \hat{f}, \varphi\rangle .$$

In other words, the Hartley transform $\hat{f}$ of a distribution of slow growth is that functional, which assigns to each φ in *S* the same number which f assigns to the Hartley transform $\hat{\varphi}$ of φ.

Since φ is in *S*, equation (234) determines uniquely the regular distribution corresponding to the ordinary Hartley transform of a locally integrable function, which is absolutely integrable over $-\infty < t < \infty$. If we consider the inverse Hartley transform, then set $\hat{f} = g$ and $\hat{\varphi} = \psi$ in (234) such that $g \in S'$ and $\psi \in S$. We obtain

$$\langle h^{-1}g,\psi\rangle=\langle g,h^{-1}\psi\rangle.$$

Theorem 9 : If f is a distribution of slow growth, then the Hartley transform $\hat{f}$ is also a distribution of slow growth. Its inverse holds true.

Proof : With linearity mentioned in (210), that is

$$\langle f,\alpha(\varphi_1+\varphi_2)\rangle=\alpha\langle f,\varphi_1\rangle+\alpha\langle f,\varphi_2\rangle,$$

we establish continuity. Let $\{\varphi_p\}_{p=1}^{\infty}$ converges in S to zero. Then by Theorem 8, Section 6.4.1, sequence $\{\hat{\varphi}_p\}_{p=1}^{\infty}$ too converges in S to zero. By (234), as $\varphi_p\in S,\hat{f}\in S'$, and as $P\to\infty,\langle\hat{f},\varphi_p\rangle=\langle f,\hat{\varphi}_p\rangle\to 0$. This shows that $\hat{f}$ is continuous. Hence, $\hat{f}$ is a continuous linear functional on S. This completes the first part of the theorem.

Now, by virtue of equation (235) and conditions mentioned above, we can prove that if $\hat{f}$ is a distribution of slow growth, then its inverse Hartley transform $f(t)$ is also the same.

6.6 Ultradistributional Hartley Transform [85]

The space Z [12 and 15] is defined to be the space of all entire functions which satisfy a set of inequalities of the type.

$$\left|(2\pi z)^k\varphi(z)\right|\le G_k\cosh(2\pi a\,|\,y\,|), \tag{236}$$

where z is a complex variable, $z=t+i(y), G_k$ and a are constants.

Theorem 10: The necessary and sufficient conditions for $\hat{\varphi}(s)$ to be in $\mathcal{D}$ with its support conained in $-a\le s\le a$, is that the inverse Hartley transform of it can be extended to an entire function that satisfies [see(236)]

$$\left|(2\pi z)^m\varphi(z)\right|\le G_m\cosh(2\pi a\,|\,y\,|), \tag{237}$$

where

$$G_m = 2\int_{-a}^{a} \left|\hat{\varphi}^{(m)}(s)\right| ds, m = 0,1,2,\ldots.. \tag{238}$$

Proof: We intend to be brief in the analysis. Let $\hat{\varphi}(s)$ is an arbitrary testing function in $\mathcal{D}$ whose support is given in the statement of the theorem. Using the relation (222), we have

$$\varphi(t) = \int_{-a}^{a} cas(2\pi st)\hat{\varphi}(s)ds. \tag{239}$$

Note that $\varphi(t)$ can be extended to an entire function over the complex z-plane such that the relation (239) reduces to

$$\varphi(z) = \int_{-a}^{a} \hat{\varphi}(s)[\cos(2\pi st) + \sin(2\pi st)]ds. \tag{240}$$

For the analyticity of $\varphi(t)$ for all finite z, we refer to Titchmarsh [107]. Integrate (240) m-times by parts, it yields

$$\left|(2\pi z)^m \varphi(z)\right| < \int_{-a}^{a} \left(\left|\sin(2\pi sz)\right| + \left|\cos(2\pi sz)\right|\right) \left|\hat{\varphi}^{(m)}(s)\right| ds. \tag{241}$$

Hence, we may obtain, from the integrand of (241),

$$\left|\sin(2\pi sz)\right| + \left|\cos(2\pi sz)\right| \leq 2\left|\cosh(2\pi sy)\right|. \tag{242}$$

Substitute relation (242) in (241), we get

$$\left|(2\pi z)^m \varphi(z)\right| \leq 2\int_{-a}^{a} \cosh(2\pi sy)\left|\hat{\varphi}^{(m)}(s)\right| ds$$

i.e.
$$\leq G_m \cosh(2\pi a|y|), \tag{243}$$

where G_m can be seen defined in (238).

Converse: For a function $\varphi(t)$, the Hartley transform is

$$\hat{\varphi}(s) = \int_{-\infty}^{\infty} cas(2\pi st)\varphi(t)dt. \tag{244}$$

Now, $\varphi(z)$ being an entire function that satisfies (237), we shift the path of integration in the z-plane to any line which is parallel to t-axis. This is justified by Cauchy's theorem and, simultaneously, for the fact that for all y in any fixed interval, $\varphi(t+iy) \to 0$ faster than any power of $1/|t|$ as

$|t| \to \infty$ as by virture of (237). Thus, we shift the path of integration from every y and as a result, form (244), we get

$$\hat{\varphi}(s) = \int_{-\infty}^{\infty} cas(2\pi s(t+iy))\varphi(t+iy)dt$$

$$= \int_{-\infty}^{\infty} [\cos(2\pi s(t+iy)) + \sin(2\pi s(t+iy))]\varphi(t+iy)dt. \quad (245)$$

Since the integral in (245) is uniformly convergent for $-\infty < s < \infty, \hat{\varphi}(s)$ possesses an arbitrary first derivative [see [107]], given by

$$\hat{\varphi}^{(1)}(s) = \int_{-\infty}^{\infty} (2\pi t)[\cos(2\pi s(t+iy)) - \sin(2\pi s(t+iy))]\varphi(t+iy)dt,$$

which, when undergoes iterated differentiation, yields

$$\hat{\varphi}^{(m)}(s) = \int_{-\infty}^{\infty} (2\pi it)^m \left[\left\{\left(\frac{1-i}{2}\right) + (-1)^m \left(\frac{1+i}{2}\right)\right\}\cos(2\pi s(t+iy))\right.$$

$$\left. + \left\{\left(\frac{1-i}{2}\right) + (-1)^{m+1}\left(\frac{1+i}{2}\right)\right\}\sin(2\pi s(t+iy))\right]\varphi(t+iy)dt. \quad (246)$$

Infinite smoothness of $\hat{\varphi}(s)$ is, therefore, justified.

The theorem is proved. In addition, one may refer to Nair and Banerji [83] and Nair [84].

Chapter 7

Wavelet Transform and Boehmians

Presumably it is a maiden venture to involve wavelet transfom with Boehmians, that we have published. We do not intend to give terminologies regarding wavelets and Boehmains. Chapter 4 is enough for details on wavelets and Chapters 3 and 5 are sufficient to learn Boehmians. For yet a brief and adequate knowledge, one may refer to [59].

7.1 Wavelet Transform and Tempered Boehmians [51]

The continuous Gabor transform (or the windowed Fourier transform) of f with respect to g is defined by [87]

$$G[f](v,t)=\tilde{f}_g(v,t)=\frac{1}{\sqrt{2\pi}}\int_{-\infty}^{\infty}f(r)\overline{g}(\tau-t)e^{ivt}d\tau \tag{247}$$

$$=\frac{1}{\sqrt{2\pi}}(f,\overline{g}_{v,t}),$$

where $f,g\in L^2(R)$ and $(f.g)$ is the inner product.

The Parseval formula of functions f,h, for the Gabor transform, is given by

$$(\tilde{f},\tilde{h})=\|g\|^2(f,h), \tag{248}$$

where

$$(\tilde{f},\tilde{h})=\int_{-\infty}^{\infty}\tilde{f}(v,t)\overline{\tilde{h}}(v,t)dvdt. \tag{249}$$

In particular, if $\|g\| = 1$, then the Gabor transform is an isometry from $L^2(R)$ into $L^2(R)$.

Theorem 1: If $[f_n / \varphi_n] \in B_I$, then the sequence $\{\hat{f}_n\}$ converge in $\mathcal{D}'$. Moreover, if $[f_n / \varphi_n] = [g_n / \gamma_n] \in B_I$, then $\{\hat{f}_n\}$ and $\{\hat{g}_n\}$ converge to same limit.

We avoid detailed proof. One can refer to, for detail analysis of the proof of this theorem, the proof of Theorem 8, Section 5.2.3

Theorem 2 : Let $F = [f_n / \varphi_n] \in B_I$ and $G = [g_n / \gamma_n] \in B_S$.

Then

(i) $(\partial F / \partial x_m)^\wedge = i x_m \hat{F}$.

(ii) $\hat{G}$ is an infinitely differentiable function

(iii) $(F * G)^\wedge = \hat{F}\hat{G}$,

(iv) $\hat{F}\hat{\varphi}_n = \hat{f}_n$.

We follow the proof of Theorem 9, Section 5.2.3, in order to complete the proof of this theorem.

Theorem 3: A distribution f is the wavelet transform of a tempered Boehmian if and only if there exits a delta sequence $\{\delta_n\}$ such that $f\delta_k$ is a tempered distribution for every $n \in N$.

Proof: If $F = [f_n / \varphi_n]$ and $f = \hat{F}$, then by (iv) of Theorem 2,

$$F\hat{\varphi}_n = \hat{F}\hat{\varphi}_n. \tag{250}$$

This proves $f\hat{\varphi}_n$ to be a tempered distribution.

Further, if $f \in \mathcal{D}'$ and $\{\delta_n\}$ is a delta sequence such that $f\hat{\delta}_n$ is a tempered distribution for every $n \in N$, then

$$F = [\left((f\hat{\delta}_n) * \delta_n\right) / (\delta_n * \delta_n)], \tag{251}$$

where $(f\hat{\delta}_n)$ is the inverse wavelet transform of $f\hat{\delta}_n$ and is a tempered distribution. Hence, so is $\left((f\hat{\delta}_n)\right)$. Then, indeed, f is a tempered Boehmian, and $\hat{F} = f$.

7.2 Wavelet Transform and Integrable Boehmians [5]

Wavelet transform is defined [Koornwinder [44], p.28] by

$$(\Phi_g f)(a,b) = \int f(x)\, \overline{g_{a,b}(x)}\, dx = F(a,b), \tag{252}$$

$f \in L^2(R^d), a \in R^*, b \in R^d$, where R is a set of real numbers, $d = 1, R^* = R/\{0\}$, and

$$\left(\Phi_g f\right) = \left(f * h_{a,0}\right)(b), \tag{253}$$

where

$$h(x) = \overline{g(-x)} \quad . \tag{254}$$

Therefore, for each $a \neq 0, \left(\Phi_g f\right)(a,b)$, where a is function of b, is the convolution product of f with dilate of a fixed h; a is a dilation parameter. The Parseval formula is [44,p.30]

$$\int f_1(x)\overline{f_2(x)}dx = \frac{1}{C_g} \int_0^\infty \int_R F_1(a,b)\, \overline{F_2(a,b)}\, a^{-d-1} da db, \tag{255}$$

where $F_1 = \varphi f_1, F_2 = \varphi f_2$, for $f_1, f_2 \in L^2(R^d)$ and C_g is the constant for admissibility condition for $0 < C_g < \infty$, for ξ almost everywhere on R^d,

$$C_g = \int_0^\infty |\hat{g}(a\xi)|^2 \frac{da}{a}. \tag{256}$$

Lemma 1: If $[f_n / \delta_n] \in B_{L_1}$, then

$$[\hat{f}_\tau]_n(v) = \int_{-\infty}^{\infty} f_n(\tilde{\tau})\overline{g}(\tau - t) e^{ivt} d\tau \tag{257}$$

converges uniformly on each compact set in R.

Proof : If (δ_n) is a delta sequence, then $(\hat{\delta}_n)$ converges uniformly on each compact set to a constant function 1. Hence, for each compact set K, for almost all $k \in K, \delta_k > 0$ on K, and

$$\left[\hat{f}_\tau\right]_n = \left(\hat{f}_\tau\right)_n \left(\frac{\hat{\delta}_k}{\hat{\delta}_k}\right) = \frac{((f_t)_n * \delta)^\wedge}{\hat{\delta}_k} = \frac{(f_t)_k * \hat{\delta}_n}{\hat{\delta}_k}. \tag{258}$$

Theorem 4: Let $F, G \in B_{L_1}$. Then

(i) $(\lambda F)^\wedge = \lambda F$ for any complex number λ and $(F+G)^\wedge = \hat{F} + \hat{G}$

(ii) $(F * G)^\wedge = \hat{F}\hat{G}$

(iii) $G[f(\tau - a)](v,t) = e^{iav}\, G[f(\tau)](v, t-a)$

(iv) $f^n(i\tau) = (-i\tau)^n = \hat{f}(\tau)$

(v) if $\hat{F} = 0$, then $F = 0$

(vi) if $\Delta - \lim_{n\to\infty} F_n = F$, then $\hat{F}_n \to \hat{F}$.

Proof may be followed from Theorem 1, Section 6.2.1 or even from Theorem 1, Section 5.3.2.

Lemma 2 : Let $f \in L_1$ and

$$f_n(\tau) = \frac{1}{\sqrt{2\pi}} \cdot \frac{1}{\|g\|^2} \int_{-\infty}^{\infty} \int_{-\infty}^{\infty} \hat{F}_g(v,t)\bar{g}(\tau - t)e^{-ivt}\, dv dt. \tag{259}$$

Then (f_n) converges to f in L_1 – norm.

Theorem 5: (Inversion theorem): Let $F \in B_{L_1}$ and

$$f_n(\tau) = \frac{1}{\sqrt{2\pi}} \cdot \frac{1}{\|g\|^2} \int_{-\infty}^{\infty} \int_{-\infty}^{\infty} \tilde{F}_g(v,t)\bar{g}(\tau - t)e^{-ivt}\, dv dt. \tag{260}$$

Then $\delta - \lim_{n\to\infty} f_n = F$ and hence, $\Delta - \lim_{n\to\infty} f_n = F$.

Proof : Let $F = [g_n / \delta_n]$ and $k \in N$. Then

$$(f_n * \delta_k)(\tau) = \int_R f_n(x-u)\delta_k(u)du$$

i.e. $$=\frac{1}{\sqrt{2\pi}}\cdot\frac{1}{\|g\|^2}\int_{-\infty}^{\infty}\int_{-\infty}^{\infty}\hat{F}_g(v,t)\overline{g}(\tau-t)e^{-i(v-u)}\delta_k(u)dt\,du\,dv,$$

i.e. $$=\frac{1}{\sqrt{2\pi}}\cdot\frac{1}{\|g\|^2}\int_{-\infty}^{\infty}\int_{-\infty}^{\infty}\hat{F}_g(v,t)\overline{g}(\tau-t)e^{-itv}F\ *\hat{\delta}_k(t)dtdv. \qquad (261)$$

Hence by Lemma 1,

$$\|f_n*\delta_k-F*\delta_k\|\to 0 \text{ as } n\to\infty.$$

The theorem is proved.

7.3 Ultra Boehmian and Wavelet Transform [57]

Wavelet transforms and ultra Boehmians have been discussed in the preceding chapter. Ultra distribution and ultra Boehmian for the wavelet transform are obtained in this section. Whenever a similarity in the proof of certain theorem is noticed, we will refer to the corresponding theorem.

Consider a complex-valued infinitely differentiable function f defined on R^N, which satisfies

$$|z|^k|f(z)|\le C_k e^{a|y|}, \forall z, \text{Im}(z)=y \qquad (262)$$

k is a non-negative integer. Such a space of all entire functions over the complex z-plane is denoted by Z, and the delta sequence is such that

(i) $\int\varphi_n=1$

(ii) $\int|z|^k|\varphi_n(z)|\le C_k e^{a|y|}, \forall k\in N.$

Considering G as Z', which satisfies (150) and (151) and the properties of the delta sequence, is called the ultraBoehmian, which is denoted by B_Z, where $f_n\in Z'$.

Theorem 6 : The sequence $\{\hat{\varphi}(v)\}_{\tau=1}^{\infty}$ converges in $\mathcal{D}$ to the limit $\hat{\varphi}_\tau$ if and only if the inverse wavelet transform $\{\varphi(\tau)\}$ converges in Z to limit φ.

Proof : Let $\{\hat{\varphi}_t\}$ converges in $\mathcal{D}$ to $\hat{\varphi}$ and all φ's are in Z. Then

$$|\varphi_n(z)| \leq \frac{a}{\pi} e^{a|y|} \sup_{v} |\hat{\varphi}_t(v)|. \tag{263}$$

Since $\{\hat{\varphi}_\tau(v)\}_{\tau=1}^{\infty}$ converges to $\hat{\varphi}_\tau$ in $\mathcal{D}$, the relation (263) is uniformly bounded. Hence, $\{\hat{\varphi}_t(v)\}$ converges in $\mathcal{D}$ to $\hat{\varphi}$. Proof is completed.

Theorem 7 : Z is a proper subspace of *S*.

Proof : If φ is in Z, then it is an infinitely smooth entire function. Its wavelet transform $\hat{\varphi}_t(v)$ is in $\mathcal{D}$, which implies that $v^m \hat{\varphi}^{(k)}(v)$ is also in $\mathcal{D}$. Hence $\hat{\varphi}^{(k)}(v)$ is again in Z. Thus, for each pair of non-negative integers *m* and *k*,

$$\left| v^m \hat{\varphi}^{(k)}(v) \right| \leq C_{mk}.$$

Hence, φ is in *S*. Finally, $\mathcal{D}$ is a part of *S* and does not intersect Z except for the zero function. Thus, it is proved that Z is a proper subspace of *S*.

Theorem 8: The space of all ultradistributions Z' contains the space of tempered distribution S'. Proof is left as an exercise.

7.4 Wavelet Transform, Fractional Integrals for Integrable Boehmians [58]

We, in this section, study the wavelet transform of Riemann-Liouville franctional integral operator on Boehmian space. Then we have obtained integrable Boehmians defined on the Boehmian space for the wavelet transform of fractional integrals.

Using the relation between the Gabor and the Fourier transform, that is,

$$G|f|(w,t) = \tilde{f}_g(v,t) = F\{f_t(\tau)\} = \hat{f}_t(v) \tag{264}$$

and

$$F(I_{\pm}^{\alpha}\varphi) = (\mp \delta x)^{-\alpha}\hat{\varphi}(x), \varphi \in L_1(a,b), \tag{265}$$

where $I_{\pm}^{\alpha}\varphi$ are Riemann-Liouville fractional integral operators, whereas the fractional integral for the Gabor transform can be expressed as

$$F(I_{\pm}^{\alpha} f_t(\tau)) = (\mp ix)^{-\alpha}\hat{f}_t(v), f \in L_1(a,b). \tag{266}$$

In other words,

$$G(I_{\pm}^{\alpha} f) = (\mp iv)^{-\alpha} \hat{f}_t(v),$$

i.e. $$G(I_{\pm}^{\alpha} f_n) = (\mp iv)^{-\alpha} (\hat{f}_t(v))_n$$

$$= (\mp iv)^{-\alpha} (\hat{f}_t)_n(v). \tag{267}$$

Theorem 9 : If $[f_n / \delta_n] \in B_{L_1}$, then the sequence

$$G(I_{\pm}^{\alpha} f_n) = (\mp iv)^{-\alpha} (\hat{f}_t)_n(v) \tag{268}$$

converges uniformly on each compact set in R.

Proof: If (δ_n) defines a delta sequence, then $(\hat{\delta}_\tau)_n$ converges uniformly on each compact set to a constant function unity. Hĕnce, $(\hat{\delta}_k) > 0$ on K and as a result, left side of (268) yields.

$$G(I_{\pm}^{\alpha} f_n) = \frac{(I_{\pm}^{\alpha} f_k)(\hat{\delta}_n)}{(\hat{\delta}_k)} \text{ on } K$$

i.e. $$= \frac{(\mp iv)^{-\alpha} (\hat{f}_t)_n (\hat{\delta}_n)}{(\hat{\delta}_k)}, \text{ by (267).}$$

This shows that the Gabor transform of fractional integral operators for an integrable Boehmian, $F = [f_n / \delta_n]$ can be expressed as the limit of the sequence $G(I_{\pm}^{\alpha} f_n)$, which (indeed) is the space of all continuous functions on R.

Chapter 8
Selected Applications

This chapter is included to illustrate some new association of Boehmians, viz. Boehmians and Torus and the Lacunary Bohmians. May be, that we presume, the torus and exchange property, involving Boehmians, are discussed for the first time through our researches.

8.1 Torus : Prologue

Not to be confused with Taurus, which is a constellation in the northern hemisphere near Orion and Airies, the second sign of zodiac. In geometry, a torus (plural tori) is a surface of revolution, generated by revolving a circle in three dimensional space about an axis coplanar with the circle, which does not touch the circle. Examples include the surfaces of doughnuts and inner tubes. A circle, rotated about a chord of it, to some context, called Torus, but it is not a common practice in mathematics. The space produced, when a circle is rotated about a chord, resembles a cushion. Torus is a Latin word for a cushion of this shape, see [S1].

A torus can be defined parametrically by, see [S2],

$$x(u,v) = (R + r\cos v)\cos u \tag{269}$$

$$y(u,v) = (R + r\cos v)\sin u \tag{270}$$

$$z(u,v) = r\sin v, \tag{271}$$

where $u, v, \in [0,2\pi]$, R is the distance from the centre of the tube to the centre of torus, and r is the radius of tube. An implicit equation, in cartesian coordinates, torus that is radially symmetric about the z-axis, is given by

$$\left(R-\sqrt{x^2+y^2}\right)^2+z^2=r^2 \tag{272}$$

and in explicit form, the equation is

$$f(x,y,z)=\left(R-\sqrt{x^2+y^2}\right)^2+z^2-r^2 \tag{273}$$

By removing the square root and rotating through 2π, equivalent to replacing x^2 by

$$(x^2/2R)-y^2 \to r=2R\sqrt{x^2+y^2}, \tag{274}$$

produces a quartic

$$(x^2+y^2+z^2+R^2-r^2)^2=4R^2(x^2+y^2). \tag{275}$$

The surface area and the interior volume of this torus is given by [S3]

$$A=4\pi^2 Rr=(2\pi r)(2\pi R) \tag{276}$$

and

$$V=2\pi^2 Rr^2=(\pi r^2)(2\pi R), \tag{277}$$

respectively. One of the most interesting surfaces in a four-dimensional space is the flat torus in three sphere. The formula, which constructs flat torus, is given by

$$[u,v]\to[\cos(u+v),\sin(u+v),\cos(u-v),\sin(u-v)]sqrt\,(2), \tag{278}$$

where u and v run from 0 to 2π. The sum of the squares of these four coordinates is unity. Hence, the object is completely contained in the hyper-sphere of radius unity, which has origin as the centre in four-space. Torus is covered with circles and when projected stereographically from the pole (0,0,0,1) on three-sphere, the image in three-space is a torus of revolution, which is covered with circles in such a manner that each goes around the torus once in each direction. For better conceptual idea, one may refer to [S4].

Topologically, torus is a closed surface, defined as the product of two circles $S^1 \times S^1$. This can be viewed as lying in C^2 and is a subset of the three-sphere S^3 of radius $\sqrt{2}$. Topological torus is often called Clifford torus. Another aspect of torus is due to Shapiro [99], by operating in Euclidean *k*- space $E_k, k \geq 2$. Following notations are used:

$$x = (x_1, x_2, ..., x_k); y = (y_1, y_2, ... y_k);$$

$$\alpha x + \beta y = (\alpha x_1 + \beta y_1, \alpha x_2 + \beta y_2, ..., \alpha x_k + \beta y_k);$$

$$(x, y) = x_1 y_1 + x_2 y_2 + ... + x_k y_k; |x| = (x, x)^{1/2}.$$

T_k designates *k*- dimensional torus $\{x: -\pi < x_j \leq \pi,\ \ j = 1,2,...,k\}, v$ will denote a point *a*, distant unity from origin, i.e., $|v| = 1$, *m* denote an integral lattice point. If $f \in L_1$ on T_k, then $\hat{f}(m)$ will denote *m*th Fourier coefficient of *f*, i.e.,

$$(2\pi)^{-k} \int_{T_k} f(x) e^{-i(m,x)} dx. \qquad (279)$$

An ordinary torus is a surface having genus one and, thus, possesses a hole. The single-hold ring torus is called an anchor ring, see Gardner [34] and references therein. The special case of a 2-torus is (sometimes) called the double torus. A fundamental result is that the phase trajectories of a Hamiltonian system, with freedom of *n*-degrees and having *n*-integrals of motion lie on an *n*-dimensional manifold, which is topologically equivalent to *n*-torus. Presumably, this could be the reason that inspired us to work on Boehmians with torus.

8.2 Mellin Transform for Boehmians on Torus [101]

The space $\beta(\Gamma)$ of Boehmians, which is general in nature on unit circle, is found in Nemzer [87,88]. He constructed a space of Boehmians which contains the space of periodic hyperfunctions and studied a subspace $\beta(T^d)$ of a space of tempered Boehmians β_J. The space $\beta(T^d)$ is considered as the space of Boehmians on torus. The investigation, which we have included in this section, involves $\beta(T^d)$ as the subspace of Boehmians on the torus for the Mellin tansform of tempered Boehmians.

Mellin transform is defined by

$$M\{f(x);s\} = \int_0^\infty f(x)x^{s-1}ds. \tag{280}$$

The relation between Mellin and the Fourier transform is

$$M\{f(x);s\} = F\{f(e^x);is\}. \tag{281}$$

The Mellin transform, in the form of the Fourier transform of $f \in S$, is

$$\tilde{f}(is) = \int_0^\infty f(e^x)e^{sx}dx. \tag{282}$$

The space of Boehmians on the torus is

$$T^d = \left\{\left(e^{ix_1},...,e^{ix_d}\right): x_j real\right\} \tag{283}$$

A function on T^d and R^d will be treated same, which is $2\pi-$periodic in each variable. The space of Boehmains on torus, $\beta(T^d)$, is defined by

$$\beta(T^d) = \{F \in \beta_j : T_{2\pi}F = F\}. \tag{284}$$

Lemma 1 : Let $F = [f_n / \varphi_n] \in B_J$. Then $F \in (T^d)$ if and only if for all $n \in N$, f_n is $2\pi-$periodic in each variable.

Proof is avoided, which is a routine work.

The Mellin coefficients for a locally integrable function *f* on T^d, are given by

$$C_k(f(ik)) = \frac{1}{(2\pi)^d}\int_{T^d} f(e^x)e^{kx}dx, k \in Z^d. \tag{285}$$

Lemma2: Let $F = [f_n / \varphi_n] \in \beta_J(T^d)$. Then the sequence $\{C_k(f_n(ik))\}_{n=1}^\infty$ converges for each *k*.

Proof : Let $k \in Z^d$. Since $\{\varphi_p(ik)\}_{p=1}^\infty$ is a delta sequence, there exists a $p \in N$ such that $\tilde{\varphi}_p(ik) \neq 0$. Hence,

$$c_k(f_n(ik)) = c_k(f_n(ik))\frac{\tilde{\varphi}_p(ik)}{\tilde{\varphi}_p(ik)}$$

$$= \frac{c_k(f_n * \varphi_p)(ik)}{\tilde{\varphi}_p(ik)}$$

i.e. $$= \frac{c_k(f_p(ik))}{\tilde{\varphi}_p(ik)}.\tilde{\varphi}_p(ik) \to \frac{c_k(f_p(ik))}{\tilde{\varphi}_p(ik)} c \quad as\ n \to \infty.$$

Theorem 1 : Let $F \in \beta_J(T^d)$. Then

$$\tilde{F}(ik) = \sum_{k \in Z^d} c_k F(ik)\delta(x-k).$$

Proof : Let $F = [f_n / \varphi_n] \in B_J(T^d)$. Then for each n,

$$\sum_{|k| \le m} c_k(f_n(ik))e^{-kx} \to f_n(e^x) \text{ in } S'(R^d) \text{ as } n \to \infty.$$

Continuity of the Mellin transform on $S'(R^d)$ implies

$$\sum_{|k| \le m} c_k(f_n(ik))\delta(x-k) \to \tilde{f}_n(ik) \text{ in } S'(R^d) \text{ as } n \to \infty$$

Hence,

$$\tilde{f}_n(ik) = \sum_{k \in z^d} c_k(f_n(ik))\delta(x-k), \quad n \in N.$$

Now,

$$\sum_{k \in z^d} c_k(f_n(ik))\delta(x-k) \to \sum_{k \in z^d} c_k(F_c(ik))\delta(x-k)$$

is in $\mathcal{D}'(R^d)$ as $n \to \infty$ and

$$\tilde{f}_n(ik) = \tilde{F}_n(ik) \text{ is in } \mathcal{D}'(R^d) \text{ as } n \to \infty.$$

Hence

$$\tilde{F}_n(ik) = \sum_{k \in z^d} c_k(F(ik))\delta(x-k).$$

One may refer to [101] for more theorems or may even refer to [102].

8.2.1 Convergence

Let U be a class of sequence on a space χ. We say $x_n \xrightarrow{U} x$ if $(x, x_1, x_2, ...)$ is in U, and is called topological if there exists a topology T for χ such that $x_n \xrightarrow{U} x$ if and only if $x_n \xrightarrow{T} x$. The space $\beta(T)$ of Boehmians on the unit circle with a convergence structure, is called $\Delta-$ convergence topological. We intend to introduce a convergence structure, called $\delta-$convergence. For the space $\beta(T^d)$, it may be noted, the $\delta-$convergence is equivalent to the $\Delta-$convergence.

Definition 1 : A sequence of function $f_n \in J$ is called convergent to $f \in J$ if there exists a polynomial p such that $(f_n - f)/p \to 0$ uniformly on R^d as $n \to \infty$.

We defing the map $L: T - \beta_J$, by

$$L(f) = \left[\frac{f * \varphi_n}{\varphi_n}\right], \tag{286}$$

where $\{\varphi_n\}_{n=1}^{\infty}$ is any fixed delta sequence.

Definition 2 : A sequence of tempered Boehmians $\{F_n\}_{n=1}^{\infty}$ is called a tempered Boehmian F that is denoted by $\delta = \lim_{n\to\infty} F_n = F$ if there exists a delta sequence $\{\varphi_n\}_{n=1}^{\infty}$ such that $F_n * \varphi_k, F * \varphi_k \in J$ for all $k, n \in N$ and for each $k \in N$, $F_n * \varphi_k \to F * \varphi_k$ in J as $n \to \infty$.

Theorem 2 : (Inversion): Let $F \in B_J(T^d)$. Then

$$F(e^k) = \delta - \lim_{n\to\infty} \sum_{|k|\le n} c_k\big(F(ik)\big)e^{-kx}. \tag{287}$$

Proof : Let $F = \left[\frac{f_n}{\varphi_n}\right] \in \beta(T^d)$ and $f_n \in C^\infty(T^d)$ for all $n \in N$. For if not, replace f_n by $f_n * \varphi_n$, then $F = \left[\frac{f_n * \varphi_n}{\varphi_n * \varphi_n}\right]$ and $f_n * \varphi_n \in C^\infty(T^d)$, $n \in N$.

Let

$$F_n(e^k) = \sum_{|k|\le n} c_k(F(ik))e^{-kx}, n = 1, 2, \ldots \tag{288}$$

Then for each p,

$$F_n \times \varphi_p = \sum_{|k|\le n} c_k(F(ik))\tilde{\varphi}_p(ik)e^{-kx}$$

$$= \sum_{|k|\le n} c_k (F * \varphi_p(ik))e^{-kx} \in J,\ n = 1,2,...$$

Also, for each $p, F * \varphi_n = f_p \in J$. Since for each p,

$$F * \varphi_p \in C^\infty(T^d),$$

$$\sum_{|k|\le n} c_k \left(F * \varphi_p(ik)\right)e^{-kx} \to F * \varphi_p$$

uniformly on R^d as $n \to \infty$. This implies that for each p, $F_n * \varphi_p \to F * \varphi_p$ in J as $n \to \infty$. As a result, hence, $\delta - \lim_{n\to\infty} F_n = F$. The proof is completed.

8.3 Hartley Transform and Lacunary Boehmians[53]

We have given definitions of Hilbert transform, the Fourier transform, the signum function, sgn(x), and the tempered distributional Hilbert transform, in the preceding sections. Nemzer [86] investigated the space of Lacunary Boehmians which has a lacunary Fourier series representation. We describe results concerning Hilbert transform for the lacunary Boehmians.

A sequence of positive integers $\{\lambda_n\}_{n=1}^{\infty}$ is called Hadamard-lacunary or simply, lacunary if there exists a constant $q > 1$ such that $\lambda_{n+1} > q\lambda_n$ for all n.

A Fourier series of the form $\sum_{n=-\infty}^{\infty} a_n \exp(i\lambda_n t)$, is called a lacunary Fourier series provided $\{\lambda_n\}_{n=1}^{\infty}$ is lacunary and $\lambda_{-n} = -\lambda_n$ for all $n \in N$. Properties of L_1 − functions, having lacunary Fourier series representation, can be seen in [42].

8.4 Main Theorems

In order to define lacunary Boehmians of the Hilbert transform, let f is a periodic function, that is, supp $f \subset (-\pi, \pi)$. Then the coefficient $\hat{f}_H(n)$ of f is

$$F(\bar{f}_H) = \hat{f}_H = -2^{-1/2} i \operatorname{sgn}(x)$$

$$\times \int_{-\pi}^{\pi} f(t)e^{-int}dt, \quad n \in Z \tag{289}$$

For

$$F = \sum \frac{i}{\sqrt{2}} sgn(x) a_n \exp(i\lambda_n t) \in \beta_\wedge,$$

we define $\hat{F} = (\lambda_n) = a_n, n \in z$. This is due to the fact that each $F \in \beta_\wedge$ has a unique representation

$$\sum_{n=-n}^{n} \frac{i}{\sqrt{2}} sgn(x) a_n \exp(i\lambda_n t) \tag{290}$$

Theorem 3 : Following statements are equivalent

(1) F is lacunary Boehmian of a Hilbert transform

(2) There exists a representation of the Boehmian $F = [f_n / \varphi_n]$ for all n, where f_n is a smooth periodic function and $F(\bar{f}_n)_H(\lambda) = 0$ for all $\lambda \notin \wedge$

(3) For every $F = [f_n / \varphi_n]$ for all n, where f_n is a smooth periodic function and $F(\bar{f}_n)_H(\lambda) = 0$.

(4) There exists a representation of Boehmian $F = [t_n / \varphi_n]$ for all n, where t_n being a trigonometric polynomial with deg. $t_n \le \lambda_n$, we have $F(\bar{t}_n)_H(\lambda) = 0$

Proof : Let us write following definitions to assist us in proving the theorem.

Definition 3 : If $F_n, F \in \beta(n = 1,2,...)$ suggests $F_n \xrightarrow{\delta} F$ (the δ-convergence for Boehmians) and there exists a delta sequence $\{\varphi_n\}_{n=1}^{\infty}$ such that for each n and $j, F_n * \varphi_j, F * \varphi_j \in C^\infty(R)$, then for each $j, F_n * \varphi_j \to F * \psi_j$ in $C^\infty(R)$ as $n \to \infty$. We write it as $\delta - \lim_{n\to\infty} F_n = F$.

Definition 4 : A sequence $F_n \xrightarrow{\Delta} F$ (the $\Delta -$convergence for Boehmians) if there exists a delta sequence $\{\varphi_n\}_{n=1}^{\infty}$ such that $(F_n - F) * \varphi_n \in C^{\infty}(R)$ for all $n \in N$, and $(F_n - F) * \varphi_n \to 0$ in $C^{\infty}(R)$ as $n \to \infty$.

Now, for a lacunary Boehmian *F*, from Definitions 3 and 4, we have $F = \delta - \lim_{p\to\infty} F_{n_p}$ for sequence of positive integer $\{n_p\}_{p=1}^{\infty}$. Hence, there will exist a delta sequence $\{\varphi_n\}_{n=1}^{\infty}$ such that

$$F_{n_p} * \varphi_m = \sum_{k=-n_p}^{n_p} \frac{i}{\sqrt{2}} a_k \hat{\varphi}_m \operatorname{sgn}(x) \exp(i\lambda_n t) \to f_m$$

in $C^{\infty}(R)$ as $p \to \infty$.

This proves that $F_{n_p} * \varphi_m$ is a periodic function hence f_m will be periodic for all *m*. Moreover,

$$\delta - \lim_{p\to\infty} F_{n_p} = [f_m / \varphi_m] \Rightarrow F = [f_m / \varphi_m].$$

Hence,

$$(\bar{f}_m)_H(\lambda) = 2\pi\, \hat{\bar{F}}_H(\lambda) \hat{\bar{\varphi}}_H(\lambda) = 0, \forall \lambda \notin \wedge.$$

This proves (i), which implies (ii). Similarly, (ii) can be proved which will imply (iii). Details are avoided due to similar line of analysis.

Further, let $F = [f_n / \psi_n]$ for all *n*, let f_n is smooth and let $\hat{f}_H(\lambda) = 0$ for all $\lambda \notin \wedge$. Then $F = [t_n / \varphi_n]$, φ_n being a delta sequence. For *F* and t_n to have usual meaning, for each $n \in N$ and $\hat{t} = 0$, we have

$$F_n = \sum_{k=-n}^{n} \frac{i}{\sqrt{2}} a_n \operatorname{sgn}(x) \exp(i\lambda_k t),$$

where $a_n = \hat{\bar{\varphi}}_m(\lambda_k) = \hat{\bar{t}}_k(\lambda_k)$ for all *k*. This proves (iii) and (iv).

We also know $\hat{\bar{\varphi}}_m(\lambda_k) \to 1/\pi$ as $m \to \infty$. We write

$$F_n * \varphi_m = \sum_{k=-n}^{n} \frac{i}{\sqrt{2}} a_k \hat{\varphi} \operatorname{sgn}(x) \exp(i\lambda_k t)$$

$$= \sum_{k=-n}^{n} \frac{i}{\sqrt{2}} \hat{\hat{t}}_m(\lambda_k)\,\mathrm{sgn}(x)\exp(i\lambda_k t).$$

Hence, for each $F_n * \varphi_m \to t_m$ in $C^\infty(R)$ as $n \to \infty$. In other words,

$$\delta - \lim_{n\to\infty} F_n = F$$

implies

$$\Delta - \lim_{n\to\infty} F_n = F$$

The theorem is completely proved.

Refernces

1. Abel, N.H. Solution de quelques problemes à l'aide d' intégraies definies, Oeuvres Complétes, Vol 1, Grondahl Christiania, Norway, pp. 16-18.
2. Antosik, P., Mikusiński, J. and Sikorski, R. Theory of Distributions : The sequential approach, Elsevier Scientific Publishing Co. Amsterdam; PWN – Polish Scientific Publishers, Warsaw, 1973.
3. Banerji, P.K. and Loonker, Deshna. A new concept of Mikusiński operators, J. Anal. Computation, **5**(2), pp.
4. Banerji, P.K. and Loonker, Deshna. On the Mellin transform of tempered Boehmians, U.P.B. Sci. Bull. Series A **62**(4) (2000), 39-48
5. Banerji, P.K., Loonker, Deshna, and Debnath, L. Wavelet transforms for integrable Boehmians, J. Math. Anal. Appl. **296** (2004), 473-478
6. Battle, G. A block spin construction of ondelettes, Commun. Math. Phys. **110**(1987), 610-615.
7. Bochner, S. Lectures on Fourier Integrals, Annals of Mathematical Studies, #42, Princeton Univ. Press, Princeton, N.J. 1959.
8. Boehme, T.K. Operational Calculus and the Finite Part of Divergent Integrals, Ph. D. Thesis, California Institute of Technology, Pasadena, Calif. 1960.

9. Boehme, T.K. The support of Mikusiński, operators, Trans. Amer. Math. Soc. **176** (1973), 319-334.

10. Bracewell, R.N. Discrete Hartley transform, J. Opt. Soc. Amer. **73** (1983), 1832-1835.

11. Bracewell, R.N. The fast Hartley transform, Proc. Leee **72** (1984), 1010-1018.

12. Bracewell, R.N. The Hartley Transform, Oxford Univ. Press, New York 1986.

13. Burzyk, J. On the convergence in the Mikusiński operational calculus, Studia Mathematica, T. **76**(1983) 313-333.

14. Cayley, A. Note on Riemann's paper, Math. Ann. **16**(1880), 81-82.

15. Chaudhary, M.S. and Thorat, S.P. On distributional Hartley transformation, Ganita Sandesh, **12** (1-2) (1998) 19-26.

16. Chui, C.K. Review of Wavelets and Operators by Yves Meyer (1992), and A friendly Guide to Wavelets, by Gerald Kaiser (1994), Bull. Amer. Math. Soc. (NS), **33**(1)(1996), 131-134.

17. Colombeau, J.F. Elementary Introduction to New Generalized Functions, North Holland Mathematics Studies # 113 Elsevier Science Publishers B.V., Amsterdam, Tokyo, 1985.

18. Colombeau, J.F. New Generalized Functions and Multiplication of Distributions, North Holland Mathematics Studies # 84 Elsevier Science Publishers B.V., Amsterdam, The Netherlands, Tokyo, 1984.

19. Daubechies, I. IEEE Trans. Information Theory **36**(1990), 961.

20. Daubechies, I. Orthonormal bases of compactly supported wavelets, Commun. Pure Appl. Math. **41**(1988), 909-996.

21. Daubechies, I. Ten Lectures on Wavelets, CBMS-NSF Regional Conference Series in Appl. Math. SIAM Pub. Pheladelphia, 1992.

22. Daubechies, I., Grossman, A. and Meyer, Y.J. Math. Phys. **27** (1986),1271.
23. Debnath, L. Internat. J. Math. Edu. Sci. Tech. 1998
24. Debnath, L. Nonlinear Instability, Chaos and Turbulence (Ed. Debnath, L. and Riahi, D.N.), Compu. Mech. Publ. (1998), 129
25. Debnath, L. Wavelet transforms and their applications, Proc. Indian Nat. Sci. Acad. (PINSA), **64**(6) A (1998), 685-713.
26. Deshna Loonker and Banerji, P.K. Cauchy representation of the Hilbert transform for integrable Boehmians, J. Anal. Comput. **3**(2) (2007), 81-86.
27. Dikshit, H.P. Multiresolution analysis and spline wavelets, A short note (unpublished), MP Bhoj Open University, Bhopal, India (Publication date not known).
28. Duffin, R.J. and Schaeffer, A.C. Trans. Amer. Math. Soc. **72**(1952), 341.
29. Erdélyi, A. Operational Calculus and Generalized Functions, Holt, Rinheart and Winston, New York, 1962.
30. Euler, L. De progressionibus transcendentibus, sev quarum termini generales algebraice dari nequent, Comm. Acad. Sci. Imperialls Sci. Petropo. **5**(1738),55.
31. Fourier, J.B.J. Théorie analytique de la chaleur, Oeuvres de Fourier, Vol.1, Firmin didot, Paris, p.508.
32. Friedlander, F.G. Introduction to the Theory of Distributions, Cambridge Univ. Press, Cambridge, 1982.
33. Gabor, D. J. Inst. Electr. Eng. London **93** (1946), 429.
34. Gardner, M. Martin Gardner's Sixth Book of Mathematical Games, pp. 15-17, Scientific American, New York Scribers, 1971.

35. Gel'fand, I.M. and Shilov, G.E. Generalized Functions, **Vol. 1-3**, Academic Press, N.Y., London 1964.

36. Goldberg, R.R. Fourier Transforms, Cambridge Univ. Press, Cambridge, 1965.

37. Grossman, A. and Morlet, J. Decomposition of Hardy functions into square integrable wavelets of constant shape, SIAM J. Math. Anal. **15** (1984), 723-736.

38. Grünwald, A.K. über "begrenzte" Derivationen and deren Anwendung, Z. Angew. Math. Phys. **12**(1867), 441-480.

39. Hartley Transform, Wikipedia, the free Encyclopedia: Source : http://en.wikipedia.org/wiki/Hartley-transform, pp. 1-4(2008 Dec.).

40. Hartley, R.V.L. A more symmetrical Fourier analysis applied to transmission problems, Proc. I.R.E. **30**(1942),144-150

41. Helson, H. Harmonic Analysis, 2nd Edn. Hindustan Book Agency, New Delhi 1995.

42. Kahane, J.P. Lacunary Taylor and Fourier series, Bull. Amer. Math. Soc. **176** (1973), 319-334.

43. Kaiser, G. A friendly Guide to Wavelets, Birkhauser, Bassel and Boston, MA 1994.

44. Koornwinder, T.H. Wavelets : An Elementary Treatment of the Theory and Applications, World Scientific, Singapore, 1993

45. Lagrange, J.L. Sur un nouvelle espece de calcul relatif a la differentiation et a l'integration des quantities variables, Oeuvres de Lagrange **Vol.3**, Gauthier-Villars, Paris, pp. 441-476 (1849).

46. Laplace, P.S. Théorié analytique des probabilities, Courcier, Paris, 1820.

47. Laurent, H. Sur le calcul des dérivées a indices quelconques, Nouv. Ann. Math. **3**(3) (1884), 240-52.

48. Lemarié, G. and Meyer, Y. Ondelettes et bases Hibertienes, Revista Mathemática Iberoaméricana, **2**(1986), 1-18.

49. Lemarié, P.G. Ondelettes á localisation expo-(3)-nentielle, J. Math. Pure Appl. **67**(1988), 227-236.

50. Letnikov, A.V. An explanation of the main concepts of the theory of differentiation of arbitrary index, Moskow Mat. Sb. **6**(1872), 413-445.

51. Loonker, D. and Banerji, P.K. Wavelet transform for the tempered Boehmians, Hadronic J. Suppl. **18** (2003), 403-410.

52. Loonker, Deshna and Banerji, P.K. Hartley transform for tempered Boehmians, Proc. Nat. Acad. Sci. India, Sect. A, **81** (Part I) (2011), 23-28.

53. Loonker, Deshna and Banerji, P.K. Hilbert transform for lacunary Boehmians, Global J. Math. Anal. **1**(1-2) (2007), 85-90.

54. Loonker, Deshna and Banerji, P.K. Kontorovich – Lebedev transformation of tempered distribution and tempered Boehmians, J.Raj. Acad. Phy. Sci. **4**(3)(2005), 165-172.

55. Loonker, Deshna and Banerji, P.K. The Cauchy representation of integrable and tempered Boehmians, Kyungpook Math. J. **47** (2007), 481-493.

56. Loonker, Deshna, Banerji, P.K. and Debnath, L. Hartley transform for integrable Boehmians, Integral Transf. Spl. Func. **21**(6) (2010), 459-464.

57. Loonker, Deshna, Banerji, P.K. and Kalla, S.L. ultradistribution and ultra-Boehmian of wavelet transform, Hadronic J. **29**(2006), 485-496.

58. Loonker, Deshna, Banerji, P.K. and Kalla, S.L. Wavelet transform of fractional integrals for integrable Boehmians, Appl. and Applied Math. **5**(1) (2010), 1-10

59. Loonker, Deshna. Boehmian spaces and wavelets, Chapter 4, pp. 40-51, in Mathematical Analysis and Applications (ed. R.C. Sharma), Allied Publisher Pvt. Ltd., New Delhi, India 2003

60. Loonker, Deshna. Distribution Spaces and Transform Analysis, Ph.D. Thesis, JNV University, Jodhpur, India, 2001.

61. Mallat, S. A theory for multiresolution signal decomposition, IEEE Trans. Patt. Mach. Intell. II (1989B), 678-

62. Mallat, S. IEEE Trans. Acoust. Speech Signal Processing **37** (1989c), 2091.

63. Mallat, S. Multiresolution approximations and wavelet orthonormal bases of $L^2(R)$, Trans. Amer. Math. Soc. **315** (1989a), 69-87.

64. Mallat, S. Multiresolution Representation Wavelets, Ph.D. Thesis, University of Pennsylvania, 1988.

65. Malvar, H.S. IEEE Trans. Acoust Speech Signal Processing (ASSP), **38** (1990b), 969.

66. Malvar, H.S. IEEE Trans. ASSP, **38** (1990a), 969.

67. Marchaud, A. Surles dérivées et sur less différences de fonctions des variables réeles, J. Math. Pures Appl. **6**(1927), 337-425.

68. Meyer, Y. Ondelettes, fonctions splines et analyses gradáees, Lecture Notes, University of Torino, Italy 1986.

69. Meyer, Y. Wavelets and Operators, Cambridge Studies in Integrated Math. **Vol. 37**, Cambridge Univ. Press, Cambridge, 1992.

70. Mikusiński, J. and Mikusiński, P. Quotients de suites et lecurs applications dans L'nalyse fonctionnelle, C.R. Acad. Sci., Paris **293**, Series I (1981), 463-464.

71. Mikusiński, J. Hiperliczby, Carcow, 1944 (Edition of 7 copies); English transl. : Hypernumbers, Univ. Calif., Santa Barbara, 1983 and Studia Math. **77** (1983), 3-16.

72. Mikusiński, J. Hypernumbers, Part I, Algebra, Studia Math. **77** (1983), 3-16.

73. Mikusiński, J. Irregular operations on distributions, Studia Math. **20** (1961), 163-169.

74. Mikusiński, J. On the value of a distribution at a point, Bull. Acad. Pol. Sci. C1. III **3** (1955), 589-591.

75. Mikusiński, J. Operational Calculus, Pergamon Press, Oxford 1959

76. Mikusiński, P. Convergence of Boehmians, Japan J. Math. **9** (1) (1983), 159-179

77. Mikusiński, P. Tempered Boehmians and ultra distributions, Proc. Amer. Math Soc. **123** (3) (1995), 813-817.

78. Mikusiński, P. The Fourier Transform of Tempered Boehmians, Fourier Analysis, Lecture Notes in Pure and Appl. Math., Marcel Dekker, New York (1944), 303-309.

79. Miller, K.S. and Ross, B. An Introduction to the Fractional Calculus and Fractional Differential Equations, John Wiley & Sons, Inc. New York, 1993

80. Morlet, J., Arens, G., Fourgeau, E. and Giard, D. J. Geophys. **47**(1982b), 222.

81. Morlet, J., Arens, G., Fourgeau, E. and Giard, D. Wave propogation and sampling theory, Part II, J. Geophys. **47**(1982a), 203-236.

82. Nair, Deepa H. and Banerji, P.K. On tempered distributional Hartley transform, Bull. Cal. Math. Soc. **102** (2) (2010), 147-152.

83. Nair, Deepa H. and Banerji, P.K. Roumieu type tempered ultradistributional Hartley transform. J. Indian Acad. Math. **31** (2) (2009), 637-648.

84. Nair, Deepa H. Distributional Spaces and their Applications to Integral Transforms, Integral Equations and Wavelet Transform, Ph. D. Thesis, J.N.V. University, Jodhpur, India. 2009.

85. Nair, Deepa H., Loonker, Deshna and Banerji, P.K. On ultradistributional Hartley transform, The Aligarh Bull. Math. **29** (1) (2010), 47-53.

86. Nemzer, D. Lacunary Boehmians, Integ. Transf. Spl. Funct. **16** (2005), 451-459

87. Nemzer, D. Periodic Boehmians II, Bull. Austral. Math. Soc. **44** (1991), 271-278

88. Nemzer, D. Periodic Boehmians, Internat. J. Math. Math. Sci. **12** (1989), 685-692.

89. Nemzer, D. The product of a function and a Boehmian, Colloq. Math. **66**(1993), 45-49.

90. Ojha, A. Wavelet Transform : An Introduction, Short Lecture Notes, pp. 1-29 (unpublished).

91. Olejniczak, K.J. The Hartley transform, in The Transform and Application Handbook, Chapter 4, 2nd edn. (ed. Poularikas, A.D.), CRC Press, LLC 2000.

92. Orton, M. Hilbert transform, Plemelj relations, and Fourier transform of distributions, SIAM J. Math. Anal. **4** (4), 656-670.

93. Rao, Geetha, S. Wavelets unconditional bases in Approximation Theory, unpublished Lecture Notes, 1-40 (date not known).

94. Resnikoff, H.L. and Wells, R.O. Wavelet Analysis, Springer – Verlag, New York, 1998.

95. Riemann, B. Versuch einer allgemein Auffassung der Integration and Differentiation, Gesammelte Werke, published posthumously, Teubner, Leipzig, pp. 355-366, 1892

96. Schwartz, L. Généérulisation de la notion de fonction, de dérivation, de transformation, de Fourier, et applications mathématieces et physiques, Annales Univ. Grenoble **21**(1945), 57-74.

97. Schwartz, L. Théorie des Distributions, I, Hermann, Paris, 1950.

98. Schwartz, L. Théorie des Distributions, II, Hermann, Paris, 1951.

99. Shapiro, V.L. Bounded generalized analytic functions on the torus, Pacific J. Math. **14** (4)(1964), 1413-1422.

100. Singh, Abhishek, Loonker, Deshna and Banerji, P.K. Fourier-Bessel transform for tempered Boehmians, J. Math. Anal. **4**(45) (2010), 2199-2210.

101. Singh, Abhishek, Loonker, Deshna and Banerji, P.K. Mellin transform for Boehmians on Torus, The Aligarh Bull. Math. **28** (1-2) (2009), 57-62.

102. Singh, Abhishek. On Boehmians and Boehmian Spaces and Bohemenian Transform Analysis, Ph.D. Thesis, J.N.V. University, Jodhpur, India, 2009.

103. Sneddon, I.N. The Use of Integral Transforms, Tata McGraw – Hill Publ. Co. New York, New Delhi 1974

104. Soboleff, S. Méthode nouvelleá réoudre le probléme de Cauchy pour les équations hyperboliques normales, Recuell Math. (Math. Sbornik) **1**(1936),39-71.

105. Sonin, N.ya. On differentiation with arbitrary index, Moscow Mat. Sb. **6**(1869), 1-38.

106. Szaz, A. Convolution multipliers and distributions, Pacific J. Math. **60**, 267-275 (1975).

107. Titchmarsh, E.C. The Theory of Functions, Oxford, 1932

108. Walker, J.S. A Primer on Wavelets and their Scientific Applications, Chapman & Hall/CRC, London, New York, 2008.

109. Wang, Z. Harmonic analysis with a real frequency function – I, Aperiodic case, Appl. Math. Comput. **9** (1981 a), 53-73.

110. Wang, Z. Harmonic analysis with a real frequency function – II, Periodic and bounded case, Appl. Math. Comput. **9** (1981b), 153-163

111. Wang, Z. Harmonic analysis with a real frequency function – III, Data sequence, Appl. Math. Comput **9** (1981 c), 245-255.

112. Watson, G.N. The Theory of Bessel Functions, 2nd Edn., Cambridge Univ. Press, London (1994).

113. Weyl, H. Bemerkungen zum Begriff des Differential- quotienten gebrochener Ordnung, Vierteljahresschr Naturforsch. Ges. Zurich, **62**(1917), 296-302.

114. Yakubovich, S.B. and Luchko, Y.F. The Hyperqeometric Approach to Integral Transforms and Convolution, Kluwer Acad. Publishers, London, 1994.

115. Zemanian, A.H. Distribution Theory and Transform Analysis, McGraw-Hill Publ. Co. New York 1965; Dover Publication, Inc. New York , 1987.

116. Zemanian, A.H. Generalized Integral Transformations, Interscience Publishers, N.Y. 1968; Republished : Dover Publications, Inc. N.Y. 1987.

[S1] Source : http://booksgoogle.com/books?

[S2] Source:http://www.geom.uiuc.edu/zoo/toptype/torus/standard/eqns.html

[S3] Source : http://mathworld.wolfram.com/torus.html

[S4] Source : http://www.geom.uiuc.edu/~banchoff/script/b3d/hypertorus.html.